The PLANNING PARTNERSHIP

The
PLANNING
PARTNERSHIP

PARTICIPANTS' VIEWS OF
URBAN RENEWAL

edited by

ZANE L. MILLER
THOMAS H. JENKINS

SAGE PUBLICATIONS
Beverly Hills / London / New Delhi

For information address:

SAGE Publications, Inc.
275 South Beverly Drive
Beverly Hills, California 90212

SAGE Publications India Pvt. Ltd.
C-236 Defence Colony
New Delhi 110 024, India

SAGE Publications Ltd
28 Banner Street
London EC1Y 8QE, England

Printed in the United States of America

Library of Congress Cataloging in Publication Data

Main entry under title:

The planning partnership

    Includes bibliographical references.
    1. Urban renewal—Ohio—Cincinnati—Citizen
participation—Addresses, essays, lectures.
2. Community and college—Ohio—Cincinnati—Addresses, essays, lectures. I. Miller, Zane L.
II. Jenkins, Thomas H. 1923-
HT177.C53P37   307.7′6   81-21532
ISBN 0-8039-1799-6   AACR2

FIRST PRINTING

Contents

DEDICATION

The Community

The late
Hon. James L. Rankin,
Ohio State Representative

The University

Dr. Hoke S. Greene,
former Vice President for Research,
University of Cincinnati

The City

The late
William Wichman,
City Manager,
Cincinnati, Ohio

The PLANNING PARTNERSHIP

Preface

This book did not have to be written. It is about an event familiar in our times, an urban renewal project in a slum neighborhood adjacent to a big city central business district, a common and frequently analyzed enterprise in American city planning practice and history. And while the contract for the project called for an academic institution, the University of Cincinnati, to provide social science and urban planning research as part of the project, the contract mentioned publications stemming from the project briefly and obliquely, in a phrase requiring the review of any such publications by the city of Cincinnati. As the planning process began, then, most participants viewed it in legal and short-range terms as yet another federally subsidized local project culminating in a plan to be approved as policy by city council, the step which presumably would terminate the active engagement of most participants with the project. The task seemed something to be done rather than a phenomenon for reflection, evaluation, and writing.

Yet, in the judgment of some participants, this project ought to be analyzed and written about. Fairly early in the planning process some of us convinced ourselves that the undertaking contained unique features and combined elements not ordinarily found together in one planning project: organized applied research involving several academic disciplines as well as planners; legally mandated citizen participation; a three-way planning partnership of community, city, and university; and efforts deliberately to establish a racially integrated inner-city neighborhood combined with an attempt to generate a community consensus in favor of such a plan. The work took place, moreover, in the challenging atmosphere of racial and urban crisis during the late 1960s. And the more complicated and controversial the process became, the deeper grew the conviction of many of us that aspects of the experience should in some way be written about and shared with others. Hence, we have put together this book.

The concern of this book is not restricted to the specifics of a particular planning project, however. To be sure, we have tried to capture the substance of that experience. But in these pages a particular planning project is used as a vehicle to raise questions of a "larger" significance and more general interest. This view required us to adopt an approach that is special, but not unprecedented, in the literature of American city planning. For example, Herbert Gans's *Urban Villagers,* Martin Meyerson and Edward Banfield's *Politics, Planning and the Public Interest,* and Alan Altshuler's *The Planning Process* all represent relatively recent studies of particular planning projects in particular cities with an interest in the utility of the findings in illuminating generic patterns and processes and in generalizing to other similar situations. We expect that some readers may be able to use this book in those ways. And we have also tried to include passages, as do these other books, depicting the drama and conflicts inherent in the politics of planning but not often candidly expressed.

Yet this book, although a case study, differs from the other three. This one deals exclusively with the planning process, and not, as does Altshuler's, for instance, with implementation. This book, moreover, by presenting essays from a variety of participants rather than the considered judgment of one scholar or a set of coauthors, offers several analytic perspectives on a process that one might otherwise take as a single reality. Also, this book includes perspectives of citizen participants in the planning process, views infrequently preserved in print, as well as analyses by academics and professionals in the planning field. Finally, the editors, in the front material and in the introductions to each of the parts of this book, attempt to point to and explicate some assumptions as well as occasional issues of larger significance either muted or implicit in the essays themselves.

We have not, however, included essays by all or even most of the participants in the planning process, for it involved a very large number of people, many of them without literary pretensions of any kind. Instead, we asked several people with different roles in the project to reflect on their participation and prepare an essay explaining and evaluating the process from their perspective. Not all of those we asked responded, and therefore we have no essays from a member of the city government or the university's central administration. But all who responded, either by submitting a manuscript or, as in one case, by submitting to a tape-recorded interview which we then transformed into an essay, are represented here. We should note, too, that as required by the contract for the project the manuscript for this book has been reviewed by Cincinnati's city government, which has consented to its publication but which has also requested us to state that *opinions in this book do not reflect those of the City of Cincinnati or any of its employees.*

As editors, we have, of course, accumulated a heavy debt of gratitude to those who helped make the book possible. We must first thank the authors who contributed essays, for they have been patient in anticipation of the appearance of their work and forbearing in the editorial liberties we have taken with it. We tried to hone their contributions to the format of this volume while preserving their meaning and tone, a process, we are sure, often frustrating for them but in which they nonetheless participated with grace.

We also want to acknowledge others at the University of Cincinnati who supported or assisted us in one way or another: Professor Edward R. Hoermann, Head of the Urban Planning Department at the time of the project, now Acting Associate Dean of the College of Design, Architecture, and Art (DAA), reviewed several early chapter drafts; Professor Jay Chatterjee, Assistant Director of the University's research and planning group and now Director of the School of Planning (DAA), helped us decide what kind of book to construct and prepared a draft of the Chronology; George Rieveschl, first as Vice President for Research and then as President of the University of Cincinnati Foundation, shared our understanding of the project's importance and urged us to do something about it in print; Professor Robert L. Carroll, Department of Sociology, who served through the project as Assistant Vice President for Research and Director of Social Science Research Institutes, has helped us in our understanding of the planning process and encouraged us in our publication efforts; W. Donald Heisel, Adjunct Professor of Political Science and Senior Research Associate, Institute for Governmental Research, whose summaries of the university social science and planning research team's reports proved useful to us in reconstructing project activities and assessing their significance; Alfred J. Tuchfarber, Associate Professor of Political Science and Director of the Institute for Policy Research, who provided word processing (typing); and Warren G. Bennis, President of the University when we conceived the idea for this book and started work toward its fruition, whose commitment to the idea of the university as an institution both in and of the city helped sustain our determination to bring out this volume.

Several people outside the university also helped in important ways. Hubert Guest, a member of Cincinnati's city planning staff during the project and now Director of City Planning, not only helped us conduct the interview which produced Chapter 5 but also provided the tape-recording equipment and covered the cost of transcribing the tape. In addition, Peter Kory, the city's Director of Urban Development during the project, kept in touch with us after departing that job and discussed his *ex post facto* view of the process with one of the editors. And we want to thank Cincinnati's current City Manager, Sylvester Murray, who expedited the city's review of the manuscript, a review not requested by him, it should be noted, but required by the terms of the contract between the city and the university.

Finally, we would like to express our sincere gratitude to all those activist residents of Cincinnati's West End, to the representatives of "the community," as they defined it, and to the other citizens and public officials whose actions and rhetoric created the project and gave the planning process its form, pace, and tang, and determined its outcome. They made possible the city's decision to undertake the project and thereby provided us the opportunity to create this book. Without them, this volume would not exist.

—*Zane L. Miller*
Thomas H. Jenkins
Cincinnati, March 1982

Contributors

LEWIS A. BAYLES, Professor and Chairman, Department of Foundations, School of Education, Atlanta University

JAYANTA CHATTERJEE, Director of the School of Planning and Professor of Planning, University of Cincinnati

HARRY C. DILLINGHAM, Associate Professor of Sociology, University of Cincinnati

JEROME R. JENKINS, Senior Research Associate, School of Planning, University of Cincinnati; Executive Director, Seven Hills Neighborhood Houses, Inc.

RICHARD W. LEWIS, Housing Management Consultant, West End Development Corporation

EDGAR J. MACK, Partner, Seasongood & Mayer, Investment Bankers; President, Indian Hills School Board

HAYDEN B. MAY, Professor and Chairman, Department of Architecture, Miami University

Introduction

This book is about how an innovative urban renewal planning process produced a particular plan for a neighborhood adjacent to a big city central business district. In 1968, the University of Cincinnati, the City of Cincinnati, and the West End Task Force, a body appointed by the city manager of Cincinnati in 1966 to assure citizen participation in planning and developing the West End, agreed jointly to plan the revitalization of a 117-acre inner-city slum known as Queensgate II.

Strictly speaking, the Queensgate II project was not the first time a three-way partnership involving a university, a city, and a community within a city had been established to revitalize an urban neighborhood. In the 1950s, for example, the University of Chicago played the central role in creating and supporting the South East Chicago Commission, which contracted with the city of Chicago and cooperated with the Hyde Park-Kenwood Community Conference to carry out an urban renewal project in Hyde Park-Kenwood, adjacent neighborhoods on Chicago's south side.[1] The similarity between this Chicago case and the Cincinnati experience is superficial, however.

First, the University of Chicago had a vital interest in the Hyde Park neighborhood. The bulk of the university's physical plant was located in Hyde Park and the institution owned a great deal of property there. Expansion of the university's physical plant was expected to take place there. Much of the university's faculty lived in Hyde Park-Kenwood and faculty children attended school there. By contrast, the University of Cincinnati was not located in Queensgate II, it owned no property there, it had no plans for expansion in the area, and its faculty neither lived nor sent their children to school in that neighborhood.

Second, the Hyde Park-Kenwood neighborhood was predominantly white and middle-class in character and closely identified socially, culturally, and politically with the university. The population of Queensgate II, on the other hand, was overwhelmingly black and poor, and socially, culturally, and politically divorced from the university.

Third, though citizen participation constituted a feature of both projects, the relationships between community representatives and the university differed sharply. In Chicago, the joint participation of the Hyde Park-Kenwood Community Conference and the University of Chicago by way of the South East Chicago Commission emerged out of a series of confrontations and negotiations between the university and the community, and developed into a strained collaborative relationship. In Cincinnati there had been no direct contact or conflict between the Queensgate II community and the university, and after the establishment of the relationship community animosity centered on the city's Department of Urban Development and the City Manager, not on the university. In Cincinnati, moreover, citizen participation was not informal but consciously and explicitly provided in the contract between the City of Cincinnati and the University of Cincinnati, a contract which the West End Task Force had helped to devise and negotiate and which it had approved by a vote of its members.

The differences between the two projects do not end there. While the University of Cincinnati Planning Team had, in effect, a dual allegiance to both the City of Cincinnati and the West End Task Force, the planners in Chicago reported only to the South East Chicago Commission, which, as the university's surrogate, had contracted with the city of Chicago to develop a plan for Hyde Park-Kenwood. In addition, the Chicago planners were not part of the university, for they were nonuniversity planners hired on an ad hoc basis by the South East Chicago Commission and quartered for convenience on the University of Chicago campus in the building that housed the Department of Geography. In Cincinnati, however, all but two members of the university's planning team were regular members of the faculty, with teaching as well as research responsibilities associated with university faculty status.

In Cincinnati, then, the university, the city, and the community participated directly as partners in planning an urban renewal project, an arrangement stemming from the historic gap separating the Hyde Park-Kenwood and Queensgate II experiences. The mid-1960s, that is, marked a great divide in the history of American city planning, in university-community and neighborhood-city relations, and in the struggle of blacks against second-class citizenship.[2] Those were the years of the invention of participatory planning, advocacy planning, and the neighborhood and community control and black power movements, and in many universities across the country some faculty members and administrators seized the moment to try and pump life into the dormant ideal of the "urban university." In one place or another people not only talked about these ideas but also tried them out in the real world of big city planning and politics. To the best of our knowledge, however, these notions merged as part of a single planning project only in Cincinnati during the Queensgate II project.

Queensgate II consists of an irregularly shaped area just west of Cincinnati's central business district (see Figure A). Sixth Street and the Millcreek

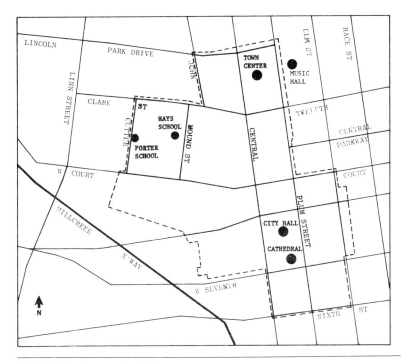

FIGURE A   Queensgate II Area

Expressway (I-75) form its southern and western borders. It stretches north to Ezzard Charles Drive (then Lincoln Park Drive). The city's Music Hall stands across the street from the area's northeast corner, and Plum Street, on which stand City Hall, St. Peter in Chains Cathedral, the Plum Street Temple, a private library, and a substation of the Cincinnati Gas and Electric Company, forms its eastern border. Also on the east lay Garfield Place, a two-block boulevard park which many regarded as a likely anchor for luxury high-rise apartments, and not far from there, but outside Queensgate II's boundaries, Shillito's department store, the city's premiere downtown shopping emporium and a corporation with a reputation for civic concern and for central business district development. Before the beginning of urban renewal in the 1950s, moreover, Queensgate II had been in the heart of the city's West End black ghetto. The fate of the area, in short, attracted the intense interest of a broad variety of local institutions and groups.

Before the partnership began planning, however, the West End Task Force had adopted general policy decisions for planning in the area which were to serve as assumptions for the partnership planning effort. The area was to be cleared and redeveloped, not rehabilitated. It was to be developed for residential and institutional land uses exclusively. Its future population could be interracial, predominantly black, or all black. And all residents as of 1967 were to be given the option of staying in the area after its redevelopment.

Under those conditions, the partnership swung into action. After the signing of the contract, the university assembled an interdisciplinary research and planning team consisting of six planners from the undergraduate planning program of the university's College of Design, Architecture, and Art, and seven social scientists from its McMicken College of Arts and Sciences. The latter were to both assist with the research on the area and consult with the planners as they developed recommendations for the Task Force. The planners also consulted with the city's Department of Urban Development, which bore responsibility for protecting the city's interest in the execution of the contract, and with the West End Task Force, which claimed authority to approve any plans for Queensgate II before such proposals went to the City Planning Commission for approval and thence to City Council for adoption as city ordinances.

Two years and $300,000 later, the City Council adopted a plan for Queensgate II. It called for the development within seven to ten years of a black or predominantly black community composed of 6,000 people living in 2,000 low-, moderate-, and middle-income housing units arranged in superblocks with small parks and courtyards and a system of one-way streets to direct traffic around housing clusters. The plan also featured: a Town Center containing shops, business and social service offices, a plaza, commercial recreation facilities, including skating and bowling, restaurants, and an Afro-American (and possibly an Appalachian) Cultural Center; a parking garage for 5,000 vehicles next to the Town Center; a pedestrian bridge running from the Town Center across Central Parkway into Music Hall; a "magnet" school and recreation complex open to students throughout the Cincinnati school district, plus three "satellite" schools for preschool to third-grade students from Queensgate II and an expanded gymnasium and athletic field at Taft High School, on the northern edge of Queensgate II, which would be open for community use; a baseball diamond for league play by youths between 16 and 25 years of age; and new industry in an area just west of Queensgate II and the West End generally.

The plan also aimed to link Queensgate II more effectively to the West End and the metropolitan area. The Town Center, for example, was expected to attract and serve people from all over the metropolitan area, as were the league baseball diamond and the expanded recreation facilities at Taft High School, the magnet school complex, and the parking garage. The pedestrian bridge was seen as a way of stimulating a flow of people from the Town Center in Queensgate II across Central Parkway to Music Hall, Washington Park across the street from Music Hall, and the Over-the-Rhine neighborhood around and to the east of Washington Park. In addition, the plan envisaged the construction of middle- and upper-income apartment housing in the Garfield Park district as a means of creating an income mix and of strengthening the racial integration of the area generally.

Whatever its merits, this plan has not yet been carried out, though someday it may be. Indeed, only two residential buildings, both high-rise apartments for low- to moderate-income residents, a structure housing the local educational television station and two radio stations, a facility for a data processing firm, and a public school building have been built in Queensgate II since 1970. The failure to carry out the plan expeditiously is not a concern of this book, however. Nor is the book concerned with the wisdom or justice of the plan or with the wisdom or justice of the partnership planning from which the plan emerged. Rather, it is concerned with an understanding of how the plan took the form it did. We have concluded, in short, that history serves us more usefully when constructed for an analysis of the "how" of things, rather than for the discovery of precedents or guidelines for current or future action.

We decided, therefore, to put together a special kind of evaluation. It is more concerned with the *process* than the product. And its concern with process centers on the how of things, by which we do not mean methodologies or even who did what to whom, though some of both appear in the book. Rather, we hope to offer a view of how the participants defined problems; how those definitions established a discourse and restricted ranges of action; how the participants acted in a context they themselves established by defining problems; and how the participants redefined problems in the context of new circumstances created by their actions. We are interested, then, in the continuing chain of problem definition, action, problem redefinition, and action that produced the Queensgate II plan.

The how of the Queensgate II plan could have been assembled and presented in several ways. It could have been assigned to the judgment of a historian, who might have sifted through the written debris of the process and interviewed participants to create a version of how things happened. That may yet be done, but such a history will be just one of thousands of possible histories of the how of the Queensgate II plan. We also recognized that many of those possible histories already existed in the heads of participants in the process who, like most human beings, try to make sense of the present and prepare for the future by constructing a memory of events from their own pasts. Those histories seemed to us worth committing to paper and pulling together. They would be interesting to us as participants in amending our own histories by revealing facts and events we did not experience, and which do not appear in the documentary record of the process. And they would be useful to a broader audience seeking to make sense of the present and to create a future by drawing on the experience of others in defining and acting on problems in a particular context and by assessing the consequences of those definitions and actions.

To this end we solicited essays from other participants in the process of making the Queensgate II plan. Not all responded, and none responded with a comprehensive account of the experience. Most wrote or talked about that part of the process in which they participated most intensely. Most of the accounts, moreover, cover events recorded also by other contributors, and frequently one account contradicts factually and interpretatively one or several other accounts. Together, then, these accounts make up an incoherent story, one which points to the elusiveness of the past—to the necessity for us to impose coherence on the past, to the sources and nature of some of the conflicts that enlivened the planning process, and to the role of those conflicts in shaping the planning process and fashioning the plan itself.

Readers should bear in mind, too, that these accounts, like all written documents, occurred after the events they record and therefore comprise constructed memories of the process. None of them is "the truth." And they should not be taken merely as part of the record of the Queensgate II planning process but also as part of another process, the process by which participants "explain" how the plan came to be and, in some cases, what happened to the plan after its adoption. In this sense, then, these accounts become part of the record of still another story, the story of what happened in Queensgate II between the signing of the contract and the present. Upon finishing the book, readers will possess information for the creation of their version of that story, the story of how people defined urban problems, how they acted upon them, and with what consequences in that recent past within which we still live.

Finally, this book may be useful in a narrower though nonetheless important way. As noted earlier, the Queensgate II plan has not been carried out, and influential people unrepresented in this volume disagree over the merits of parts or all of the plan. Something remains to be done in Queensgate II, and the argument over what continues. This book therefore can provide both information and perspectives useful in the continuing discourse about the fate of Queensgate II and, by extension, of the city and metropolis of which it forms a part.

To these accounts we have added an abstract of the plan to provide readers with the official rules of the game (see Appendix). We have also prepared a chronology to assist readers in keeping the facts straight and to provide a sequence of events with which any attempt to construct a history necessarily begins. Finally, we have reproduced a history of the Queensgate II neighborhood that was prepared as part of the planning process (Chapter 2). It contains not only one view of the history of the neighborhood, but also aspects of the social and institutional history of Cincinnati for those unfamiliar with this terrain. This document also played an important role in shaping the general goals and objectives of the plan and is therefore a significant part of the how of the Queensgate II planning experience.

NOTES

1. Peter H. Rossi and Robert A. Dentler, *The Politics of Urban Renewal: The Chicago Findings* (New York: Free Press, 1961).

2. For a fuller exposition of the differences between the years before and after the mid-1960s see Zane L. Miller, *Suburb: Neighborhood and Community in Forest Park, Ohio, 1935-1976* (Knoxville: University of Tennessee Press, 1981).

Chronology

1948	City adopts the Cincinnati metropolitan master plan, recommending construction of Mill Creek Expressway through the West End and slating area for demolition and redevelopment.
1959	City adopts West End redevelopment plan designating that part of the West End northeast of the expressway (Queensgate II) for predominantly residential land use.
November 1961	Queensgate I (nonresidential, other side of expressway) approved by federal authorities.
September 1963	William Wichman becomes City Manager.
June 1964	City considers demolition of two West End residential blocks to provide more parking for Crosley Field (then the home of the Cincinnati Reds) patrons.
April 1965	West End Community Council demands immediate redevelopment of Queensgate II and creation of a Task Force of West End citizens and city officials to plan the project.
August 1965	Queensgate II Community Club established.
January 1966	West End Community Council protests any plan to build a new stadium for professional baseball and football in West End.
April 1966	City Manager Wichman establishes West End Task Force, Edgar J. Mack, Chairman.
November 1966	William Mallory elected to state legislature.
Summer 1967	Racial disorders in Cincinnati.
July 1967	West End Task Force recommends preparation of federal urban renewal planning grant application for Queensgate II. West

	End Task Force approves planning policy decisions for Queensgate II prepared by City Planning Commission staff aids for the Task Force.
August 1967	City Council approves resolution authorizing federal urban renewal application for Queensgate II.
February 1968	Hubert Guest, City Planning Commission staff aid to the West End Task Force, announces pending retention of a University of Cincinnati team to assist in planning Queensgate II.
April 1968	City Council approves the policy decisions of the West End Task Force for Queensgate II.
	City Manager Wichman resigns.
May 1968	West End Task Force recommends approval of the proposed contract with the university.
	Task Force Chairman Mack presents Thomas H. Jenkins as possible director of the project for the university.
	Mayor Eugene Ruehlman leads a Cincinnati delegation to Washington, D.C., for a meeting with a representative of the Department of Housing and Urban Development to insist on reservation of funds for Queensgate II planning and redevelopment.
	One week later, the West End Task Force holds a press conference to announce the signing of Queensgate II planning contract with the University of Cincinnati and to formally introduce Thomas H. Jenkins as director of the university's planning and research team.
June 1968	Richard Krabach appointed as City Manager.
July 1968	Jayanta Chatterjee presents university team's work program to the West End Task Force. Unanimously approved the next day.
August 1968	Thomas H. Jenkins submits to City Manager "An Interim Improvement Program for Queensgate II" to remedy some immediate problems in the area.
September 1968	West End Task Force approves the "Interim Improvement Program."
	West End Task Force recommends that the university team investigate the feasibility of a local shopping center for Queensgate II.
October 1968	City Planning Commission approves West End Task Force's policy decision for Queensgate II.
	Some West End Task Force members complain of lack of communication with the university team and Task Force adopts resolution requesting the university's team to uti-

	lize the Task Force's policy decisions in developing plans for Queensgate II.
December 1968	Thomas H. Jenkins and Hayden B. May present to the West End Task Force the university team's "Alternative Development Goals" for Queensgate II. Some Task Force members object to the extensive use of "jargon," which they found difficult to understand.
	Task Force votes on "Alternative Redevelopment Goals," making each decision by unanimous vote and calling for mixed-income housing for families and individuals.
January 1969	West End Task Force discusses role of small business in Queensgate II planning and redevelopment.
March 1969	Peter Kory becomes director of the city's Department of Urban Development.
May 1969	Thomas H. Jenkins and Jayanta Chatterjee subjected to a surprise "confrontation" session with black leaders.
	University team presents land use and transportation proposals to West End Task Force.
	West End Development Corporation (WEDCO) chartered by state of Ohio.
June 1969	West End Task Force approves unanimously the university team's land use proposal.
July 1969	West End Task Force adopts housing proposal (alternative Y).
August 1969	West End Task Force approves transportation proposal.
July 1970	West End Task Force adopts WEDCO's amendments to housing plan for Queensgate II.
August 1970	WEDCO and Mid-City Development, Inc., sign an agreement as joint developers of Queensgate II.
	University completes Queensgate II work activities.
October 1970	City Council unanimously approves Queensgate II plan.
September 1971	WEDCO-Mid-City sign contract with city for development of first phase of the Queensgate II project.

PART I

Context:

CONCEPTUAL AND HISTORICAL PERSPECTIVES

It is important for an appreciation of the significance and larger meaning of the Queensgate II urban renewal project and planning process to see them in context. By context we mean both the conceptual trends in urban planning during the 1960s and the conditions, which, as understood by the participants, had developed historically, that both provided the major components of the problems as defined by the participants and set parameters for planning solutions. In this section, Thomas H. Jenkins, director of the university planning and research team during the project, and Zane L. Miller, the historian who served as a social science consultant in the planning process, introduce these conceptual and historical perspectives.

Jenkins presents an analysis of American city planning ideas and programs and the broad intellectual climate that affected urban planning and renewal during the 1960s in almost any city in the country, including Cincinnati. As Jenkins sees it, the general directions of these ideas and practices were away from a technical and neutral approach toward people-centered planning, and away from prescriptive blueprinting toward coordinative and policy planning. Jenkins argues, too, that these new directions in planning were greatly influenced by positions taken by politically active professional planners, by notions of social responsibility for professionals and academics advocated by intellectuals with university affiliations, by the new radical political ideas of racial minorities and others, and by social changes flowing from this intellectual ferment. But Jenkins's principal contention is that the 1960s constituted a period of passage from things as they were in planning and urban development to what they have since become. He thus characterizes the decade as a turning point between the past and the present, a period of virulent debate about new ideas accompanied by new kinds of urban social disruptions and transformations of federal and local governmental programs. He also sets down his view of how these developments affected the planning process for Queensgate II.

Miller's history of Queensgate II depicts it as an integral part of the history of the form and structure of the Cincinnati metropolitan area from the early nineteenth century into the 1960s. Unlike Jenkins, however, whose essay is concerned with ideas and programs in the 1960s, Miller concentrates on Queensgate II as a place of change and transition in people, institutions, and land uses over a long period of time. According to Miller's chapter, Queensgate II changed from a peripheral urban settlement in the early nineteenth century to a diverse area of mixed land uses and people in the expanding city of the mid-nineteenth century, to an inner-city but nonetheless lively slum in the late nineteenth and early twentieth centuries, and finally, in the 1950s and 1960s, into an isolated and dying neighborhood. It should be noted, too, that Miller wrote this essay as a research report for use in the Queensgate II planning process, and it appears here as submitted then. It proved to be the most important contribution to the project by a university team member, for its analysis of the plight of Queensgate II in the 1960s established the ground of understanding for critical decisions, and its argument about what Queensgate II had been and by implication might become set the tone for both social and physical development policy considerations.

CHAPTER 1

The 1960s—A Watershed of Urban Planning and Renewal

Thomas H. Jenkins

The decade of the 1960s was a turning point for a number of major planning ideas and urban development policies and programs. It was a decade during which some developments came to maturity and went into decline, while others that then first came into being became strong in the 1970s. Also, some long-standing planning theories and program policies underwent metamorphoses during the 1960s and emerged in new forms in the early 1970s. In these ways the 1960s proved to be a watershed for community planning and urban development in the United States. The planning work and decision-making processes in Cincinnati and in the Queensgate II urban renewal project generally reflected these changes and themes.

This chapter will review developments in American city planning and urban improvement programs during the 1960s, as a framework in which to make the Queensgate II project more understandable on its own and more meaningful in relation to other urban renewal projects that took place in the United States during the same general period—and since. An understanding of the Queensgate II project, in turn, may be helpful in comprehending the impact of national ideas, trends, and policies on local communities during the 1960s and 1970s.

The 1960s may be characterized as a pivotal period, a great divide between patterns and developments in planning and renewal issuing from the 1930s, 1940s and 1950s, on one hand, and ended, changed, or maintained in the 1960s, and inherited by the 1970s, on the other. Figure 1.1 outlines the pattern. The present chapter consists of a discussion of some of these trends, how they affected Queensgate II planning and were reflected in the features and policies of that project. What follows is a discussion of changes in a wide range of ideas and programs, and then a focused look at four that were especially characteristic of the 1960s and 1970s, and that were pertinent to the Queensgate II project.

The overall tone of urban planning and renewal crystallizing from about 1960 to 1970 may be summed up in two themes of change: from the rational and technical approaches to more human approaches, and from comprehensive planning to coordinative planning. The movement toward a more humanizing approach to urban development, steadily departing from an earlier preoccupation with land use, zoning, and economic objectives, was part of a larger pattern of social and public policy change. The 1960s was the age of the antipoverty program, with its Head Start projects and day-care centers, the Model Cities program, community health planning and clinics, and community-controlled schools. These were quite different from the concerns of the 1930s, 1940s, and even the 1950s, which centered on such things as physical redevelopment, public housing, municipal tax bases and duplicates, real estate investment, the highest (economic) use of urban land, compatibility of land use activities, and the logic and order of the master plan. In the late 1950s and through the 1960s, however, the mood shifted more toward an interest in people and their problems, and away from land and property. There is one view that all of this has been part of a more general decline in rational values in the United States in recent times and a corresponding change in the bureaucracies that have sustained these values[1]—a change noted at the tail end of the 1960s and the beginning of the 1970s. The Queensgate II plan was mainly a people-oriented plan, as will be seen later. And often, in the process of arriving at it, some of the older logico-rational canons of planning were eschewed. In turn, many of the new ideas and programs were reflected in Cincinnati and Queensgate II planning (see Table 1.1).

The tendency to move from comprehensive planning to coordinative planning has gone so far in Cincinnati, Boston, and other cities that it has replaced master planning and other forms of centralized, comprehensive

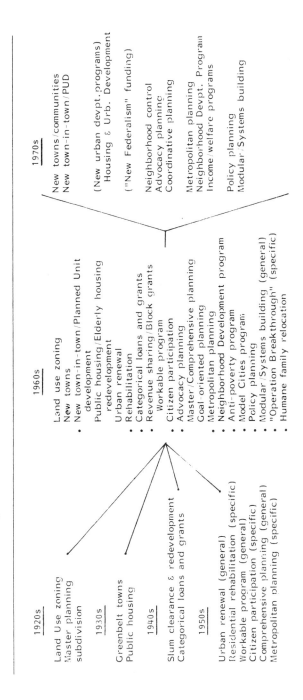

1920s

Land Use zoning
Master planning
subdivision

1930s

Greenbelt towns
Public housing

1940s

Slum clearance & redevelopment
Categorical loans and grants

1950s

Urban renewal (general)
Residential rehabilitation (specific)
Workable program (general)
Citizen participation (specific)
Comprehensive planning (general)
Metropolitan planning (specific)

1960s

Land use zoning
New towns
• New town-in-town/Planned Unit
  development
Public housing/Elderly housing
  redevelopment
Urban renewal
• Rehabilitation
• Categorical loans and grants
• Revenue sharing/Block grants
  Workable program
• Citizen participation
• Advocacy planning
  Master/Comprehensive planning
  Goal-oriented planning
  Metropolitan planning
• Neighborhood Development program
• Anti-poverty program
• Model Cities program
• Policy planning
• Modular/Systems building (general)
• "Operation Breakthrough" (specific)
• Humane family relocation

1970s

New towns/communities
New town-in-town/PUD

(New urban devpt. programs)
Housing & Urb. Development

("New Federalism" funding)

Neighborhood control
Advocacy planning
Coordinative planning

Metropolitan planning
Neighborhood Devpt. Program
Income/welfare programs

Policy planning
Modular/Systems building

FIGURE 1.1   1960s Decade as Watershed of Planning Ideas, Approaches, and Programs

• Signifies that program was inherited and retained from previous decades, and/or carried into 1970s, possibly in modified form.

TABLE 1.1 Planning Ideas and Program Developments in the "Watershed" 1960s and Their Relations to Cincinnati and Queensgate II

American Planning Ideas and Developments*	Affected	
	Cincinnati	Queensgate II
new town-in town	Yes. Re Queensgate II	Yes. Urban renewal director proposed "New York-type" concentration of 3000-5000 dwelling-unit scheme (versus 2000-unit plan chosen).
residential rehabilitation	Yes. "Operation Rehab" was largest program in the United States.	No. QII was a clearance and redevelopment project.
federal block grants: demise of categorical loans & grants	Yes. Revenue sharing	Yes. Per city administration discretion. (Only slight before 1970s.)
citizen participation	Yes. Urban riots. CBD Working Review Committee. West End Task Force.	Yes. West End Task Force and City-University of Cincinnati contract. West End Community Council's West End Development Corp. role as codeveloper of Stage I project. Queensgate II Club plan influence. Father Sicking and Black Turks.
advocacy planning	Yes. City Planning Commission staff sympathies with West End residents in planning options. Seven Hills Neighborhood Houses, Inc.. "community development" activist role. "Connection" UC student community service project. VISTA/American Institute of Architects community service project.	Yes. Certain professionals on West End Task Force. Planning staff and community professionals re West End health clinic for QII and rest of area.

metropolitan planning	City manager study called for city-county government. OKI regional transportation, housing studies. Public housing efforts to metropolitanize distribution. Indianapolis "Univgov" posing pressure by example.	Yes. Concentrate resources on inner city QII as part of resistance to City Manager study.
neighborhood development program	Yes. City complied with new HUD renewal fiscal spending. CBD, QII. Avondale-Corryville, among affected projects.	Yes. One of affected projects.
antipoverty program	Yes. Community Action Commission as CAP agency.	No. Except for relation to Model Cities program.
policy planning	Yes. CBD Working Review Committee, West End Task Force.	Yes. QII Plan and rationale.
modular/systems building	Yes. Slightly. Renewal director sought to use Built Environment, Inc., for housing projects. City applied to be one of few Operation Breakthrough cities of Nixon Administration program.	Yes. QII conceived as pilot project for Built Environment work, and part of site for proposed Breakthrough project, as part of new town-in town idea.
humane relocation	Yes. City policy stance reaction to criticism of "poorly handled" Queensgate I project, like relocation problems of 40s, 50s.	Yes. Strong QII assurances against "spectre of Queensgate I" and reaction to relocation of 40s and 50s.

*See programs and developments, 1970s column, in Figure 1.1.

planning that originate in and are maintained by a city's planning department. Emerging from the 1960s is a clear acknowledgment by many city
planning professionals and officials that their most effective work may be in
coordinating the separate planning efforts of several major structures and
units in the urban community. Hospitals and large medical centers, health
departments, boards of education, transit companies, utility companies, and
large police departments each do their own planning through large staffs,
decision-making organizations, and budgeting systems. These modern urban facts of life became pretty clear in the course of planning an urban
renewal program such as Queensgate II—especially in the seeking of planning agreements and program commitments from these various agencies and
institutions.

The demise of comprehensive and centralized planning and development
is not too surprising, because a number of planners, in the height of the urban
renewal movement, argued the virtues of subunit project planning. Spokesmen such as Herbert Gans, one of the most influential writers on American
planning, began proposing in the late 1950s that *goal*-oriented planning was
more relevant to people than physical planning programs or *community-
wide*-oriented planning.[2] Goal-oriented planning is that type of planning
geared to people and their values, aspirations, and objectives, and not to the
interests and welfare of an abstract, amorphous overall "community." This
idea had taken firm footing in the 1960s, and inasmuch as it gained momentum in the planning schools, it continued through the 1970s. Gan's paper on
goal-oriented planning was published ten years later, in 1968,[3] the year the
University of Cincinnati team started its Queensgate II planning work. This
had an effect on our attitude toward Queensgate II planning and toward the
community.

During the late 1950s and the 1960s there were many others who either
shared Gans's ideas on goal orientation or his corollary criticism of comprehensive planning. On the latter question, Alan Altshuler went so far as to
argue in the mid-1960s that the approach of city planners should in fact be
piecemeal rather than comprehensive.[4] The idea was to limit the ever larger
governmental scope that planners tended to propose in pursuit of some
"comprehensive" purpose, a purpose that Altshuler felt lost sight of the
day-to-day and down-to-earth needs of people and their communities. He
reasoned further that in order to acknowledge the government (rather than
the planners) as the rightful and ultimate public interest decision maker,
planning should be coordinative and untrammeled by preconceived *long-*

range future commitments.[5] These kinds of ideas were sometimes the subject of discussion among laymen and community development professionals representing the West End and the Queensgate II community in Cincinnati, and occasionally between the planners and members of the community. Their view was that completion of planning for development of Queensgate II should not be swallowed up and lost in the comprehensive consideration of city or metropolitan plans and programs.

Within the framework of these two tendencies toward more human approaches and coordinative planning, several specific elements and ideas were prominent in and characteristic of the 1960s. These included citizen participation, advocacy planning, social planning (antipoverty and more humane family relocation programs), coordination of physical and social planning (Model Cities program), stepped-up emphasis on residential rehabilitation, and policy planning. There were also several other elements, not reviewed in this chapter, that should be mentioned because they either contradicted one of the two themes of change described here, or modified previously existing programs or planning ideas. Specifically, they were: a new form of metropolitan planning and government as a *county* jurisdiction,[6] the decline of urban renewal total projects and categorical grants (annual block grants in the Neighborhood Development Program), and New Towns and Newtown-Intowns (later, the New Communities Act) in the 1960s. Finally, there was a new development outside of the two themes of change that affected Queensgate II planning: the popularization, within architectural, planning, building, and housing circles, of the modular or "component" system of construction (that is, factory-built housing) and encouragement of it by the federal government (the Nixon Administration's Operation Breakthrough; see Figure 1.1).

All of these elements of urban planning and renewal in the 1960s reached into the 1970s and were especially relevant to Cincinnati and Queensgate II. The four most relevant to Cincinnati-Queensgate II, and to many other local planning and renewal efforts in the United States, were: citizen participation, advocacy planning, coordination of physical and social planning, and policy planning and research. These deserve more detailed discussion.

Participation normally is carried out through organization. Civic organizations have been actively involved in urban planning for a long time. The Citizens Housing Council of New York and a variety of citizens' planning associations flourished in Los Angeles, Pittsburgh, Denver, Philadelphia, Milwaukee, Buffalo, Syracuse, and other cities both during and following

World War II. In a golden era of planning, private citizens' organizations such as the City Club of Chicago and the Regional Plan Association of New York were promoting projects of civic design and regional planning in the 1920s and 1930s.[7] As long ago as 1907 the City Club of New York was active in planning activities, and the Civic League of St. Louis produced "A City Plan for St.Louis."[8] During the 1940s and 1950s, a variety of community councils, citizens' housing associations, and community health councils engaged primarily in housing and social planning programs.[9]

The participation of all such groups, until recently, however, has been unofficial, and sometimes peripheral. Their bid for more direct participation in planning did not begin until late in the 1950s and was not institutionalized in governmental and community programs until the 1960s. This new form of citizen participation appeared in cities around the country in the 1960s and 1970s, and it played an important role in the Queensgate II project.

The Workable Program of the 1954 Housing Act called for citizen partici-pation as one of seven requirements for which a city had to give annual evidence in order to continue receiving federal funds for assistance in its urban programs. It was not until the 1960s, however, that this requirement was seen as being concerned also about "the *kind* of citizen participation a community is able to achieve."[10] It was part of a general political pressure for more genuine involvement, to replace the impotent role of token participa-tion represented by a mere advisory role. But to many citizens and planners, by the middle of the 1960s even this requirement had devolved into a farce. It seemed merely a call for "citizen ratification" of official decisions[11] or a shallow game of listing names and numbers to dress up a city's Workable Program report.

Thus, before the Workable Program, and before 1960—except for such projects as Saul Alinsky's Back-of-the-Yards and Chicago's Hyde Park re-newal program—most citizen participation in urban planning had been either public educational, civic promotion and betterment, legislative, pub-lic persuasion, critical, or advisory. Most of these groups stood outside of the structure of power of urban development, that is, outside of the decision-making process of planning and implementation and outside of the relevant political decision-making process.[12] The struggle that ensued over the rec-ognition of this fact by citizens and their allies from the professions was carried on in such a way—both inside and outside of planning—as to influence citizen participation in planning in the 1970s and beyond. The history of citizen participation in planning, then, has developed in several phases, going from token participation to a more genuine role.[13]

The more direct forms of citizen participation were characteristic of the 1960s, and this kind of community involvement was associated in many places in America with a more aggressive brand of participation. The prevailing climate of community participation, and of minority or underdog political activity, was *militancy*. About a year after President John F. Kennedy's inauguration in 1961, a steadily rising aggressive mood made social conflict and militant action marked characteristics of the decade. Manifestations of this mood included: stronger union demands and strikes by old-line labor organizations; an intensification of the civil rights movement; student campus rebellions; the unionization of major league baseball players and other professional athletes; more active pressure for a say in management by previously organized professions, such as public school teachers; the advent of bargaining efforts by such a previously *un*organized professional group as university professors; the anti-Vietnam war protest movement; the incipient stages of the women's liberation movement; and the widely scattered urban rebellions of black communities and militant black organizations.[14]

This militant mood was also present in Cincinnati, including the West End and Queensgate II, in the mid- and late 1960s. It was manifested most dramatically in Cincinnati's urban riots in the summer of 1967[15] and in the University of Cincinnati campus disturbances of 1969. Shades of this militant mood were reflected in the social and political processes of Queensgate II planning. These included a desire for neighborhood identity and self-determination, that is, community control.

The decentralization of municipal government (and national government) that took place in the late 1960s, combined with militancy, was a natural climate for the quest for community control. In his glossary of urban terms, Charles Abrams defines and characterizes community (or neighborhood) control:

> The control of a public facility or service by a local community. . . . Demands for neighborhood control go beyond mere "citizen participation" or "consultation" to the power to allocate resources, set policy, hire and fire, and plan and run programs.[16]

The community control movement burgeoning in the 1970s was a logical outgrowth of the emphasis on citizen participation pushed and provided for in the antipoverty ("maximum feasible participation") and Model Cities programs in the late 1960s. It was part of the drive for "participatory democracy" that grew to a peak in the late 1960s, accompanied by administrative decentralization of public services.[17] This trend occurred in Cincinnati, too, as the city formed several neighborhood task forces for social and community planning, and in 1969 City Manager Krabach "proposed a new plan of operations to bring City Hall closer to the people."[18]

The nature of this new movement, and its 1960s "watershed" effect portending the 1970s, was summed up in one paragraph a few years ago by George Frederickson:

> A new slogan—participatory democracy—came into use in the early 1960s. A new practice—maximum feasible participation of the poor—began in the mid-1960s as part of the anti-poverty program. A new demand—community control—was heard as the 1960's ended.[19]

This social movement was particularly relevant to black communities, where most of the action took place, and where most of the issues had been quartered, agitated, and exercised.[20] Indeed, Alan Altshuler saw the cry for "community control" largely as a "black demand for participation in large American cities."[21] This community "demand" (Frederickson and Altshuler) in the late 1960s and early 1970s for more "power" (Abrams) was an aggressive form of citizen participation that was fostered in turn by a larger social and political envelope of militancy in the general community. And nowhere was this more apparent than in the urban black community.

The predominantly black Queensgate II urban renewal area and the West End exhibited mildly aggressive forms of the community control spirit in social action. Our later discussion of the West End Task Force details several instances of community control efforts in a two-year period, relating to urban renewal planning, health planning and service, physical urban development control, and family relocation. More aggressive participation and increasing control have thus been the recent patterns of the community role.

These more militant, more radical forms of *citizen* participation have been accompanied by new *planning* philosophies and directions: advocacy planning, coordination of physical and social planning, and policy planning and research.

Advocacy planning as an idea and movement in the planning field developed slowly over several decades, and underwent such a dramatic change of form in the 1960s that in its new shape it continued as a prevailing force in the 1970s, with some survival in the 1980s. Its precursors start back at least as far as the 1930s, stretching forward almost thirty years up to the early 1960s: Saul Alinsky's Industrial Areas Foundation operating out of Chicago in disadvantaged communities in various cities, such as Rochester, New

York City, and Chicago, from the 1930s to the 1970s; the Hyde Park-Kenwood Community Conference in Chicago,[22] the Southeast Chicago Commission at the University of Chicago, and similar groups in other cities, in the 1950s; and the Architects Renewal Committee for Harlem (ARCH) in New York[23] and United Planning Aid (UPA) operating in Cambridge and Boston, headquartered in Cambridge.[24] These were some of the sociological and political antecedents around the country signaling a new kind of planning.

Some writers have circumscribed, questioned, and enlarged upon the concept of this more political type of planning. On the one hand, Keyes and Teitcher have cautioned about the "limitations of advocacy planning";[25] while on the other, Friedmann has seen advocacy as a constituent part of his larger and later concept of "transactive planning."[26] That is, Friedmann sees advocacy as a transactive "style" of planning by which knowledge is translated into action through interpersonal relations and communication between professional planners and their clients.

Advocacy planning clearly is neither technically nor primarily a part of citizen participation, although intimately associated with it. It is a movement among *professionals:* planners working with and for (underprivileged) citizens as clients. The principal change in the profession in the 1960s was that this prescribed a politically activist role for the planner, as opposed to the traditional role of dispassionate expert and neutral and rational technician.

Saul Alinsky's brainy, militant, community consultant was a specific prototype of this kind of planner.[27] A more general prototype, however, is found in the recent aggressive community organizer and social planner: "a new occupational grouping of people who meddle, organize, and plan with and on behalf of others . . . Their activities are not entirely new. [But] their professionalization is a recent phenomenon."[28] These respectively are the forerunners and seedbed of the "change agent" and the advocate planner. Social conflict, negotiations in power, and identification with the underdog were the cornerstones of their approaches.

"Advocacy" planning got its name and start as a popular movement among planners shortly after the middle of the 1960s. In 1965 Paul Davidoff made the first articulate statement that proved to be effective in influencing city planners to help the "have nots" plan their own neighborhoods instead of planning merely through the central establishment at city hall which, although it purported to be serving the "public interest," served in reality the interests of the "haves."[29] To the extent that professional planners did assume this role of advocate for special interests rather than working merely for a generalized interest, it was believed by advocacy proponents, it would promote a more equitable pluralistic democracy in urban America. An attorney as well as planner, Davidoff used the advocate lawyer in a trial court as a guiding model for this concept of a new role and function for the planner. Instead of planning "comprehensively" for the total community, this new type of planner would

> be responsible to his client and would seek to express his client's views. This does not mean that the planner could not seek to persuade his client. In some situations persuasion might not be necessary, for the planner would have sought out an employer with whom he shared common views about desired social conditions and the means toward them.[30]

After Davidoff's clarion essay, the movement expanded and the literature multiplied.[31] One of the central purposes of advocacy planning, as practiced, was to help blacks, other minorities, and poor whites achieve some balance of power, influence, and benefit with middle- and upper-income whites whose interests generally were believed to be favored by central, established government.

The term *advocate planner* was something of a misnomer at the beginning of the 1960s because most of the earliest advocate planners so designated were *architects*. Prime examples are Richard Hatch, founder of the Architects' Renewal Committee for operating ARCH in Harlem in New York, and Robert Goodman, a founder of United Planning Aid (UPA), a neighborhood assistance program in the Boston-Cambridge area.[32]

Accordingly, the earliest major advocacy projects and programs were mainly composed of, led, or staffed by architects. The most noticeable manifestation of these developments around the mid-sixties was the proliferation of community design centers (CDCs) preceded by, and later linked to, an advocacy-type unit of the federal government's VISTA[33] program, whereby the American Institute of Architects supplied the personnel of young graduate or early-career architects, while VISTA provided the funds for "community-directed" projects in central-city urban ghettos and other disadvantaged neighborhoods.[34]

Cincinnati's counterparts of the nationwide programs of AIA-VISTA/ CDCs and of Harvard's UFS neighborhood assistance programs, respectively, were (a) limited AIA-VISTA projects in Northside and one or two other communities, and (b) the University of Cincinnati's CONNECTION project, operating through the undergraduate Department of Architecture with local AIA (American Institute of Architects) assistance; and a partially successful CHART project (sensitivity training for change agents), and a partially unfulfilled general program, both of the university's then independent Graduate Department of Community Planning. CONNECTION was similar to Harvard's United Field Service (UFS), providing academically credited student architectural design and planning services, free of charge, to several communities.

Advocacy planning for the Queensgate II urban renewal project took several unorthodox forms and came from various places, such as the City Planning Commission, the university planning team, the Seven Hills Neighborhood Houses, and the medical school's Kettering Laboratory at the university. These activities are described later in connection with the West End

A second major development in the planning field accompanying stronger citizen participation was the deliberate coordination of physical and social planning. The Model Cities program represented an explicit federal policy to attempt the integration of these two forms of planning in the core areas of central cities, in a massive, publicized, and nationally funded undertaking. Congress gave it both authorization and appropriations in 1966—squarely in the middle of the sixties. It was the first official, large-scale, conscious break point in what up to that time had been, on one hand, a longitudinal succession of federal programs for *physical* planning and development, and, on the other, a nearly parallel separate succession of federal *social* planning programs. Figure 1.2 illustrates, historically, how the Model Cities program became in the mid-sixties the preeminent junctioning of federal physical and social planning programs.

For about thirty years—from the early 1930s until the early 1960s—the federal government produced a number of major urban programs, at some points in ad hoc and incremental fashion. Most of the physical programs dealt with private and public housing, land acquisition, slum clearance, residential rehabilitation, public works, and public facilities. The series of social programs involved labor, education, health, and poverty. For most of the period from about 1933 to about 1963, there was no tangible sign of coordination of any of the many physical programs with any of the variety of social programs.[35]

While the various pre-1964 federal health, education, and employment programs may have been dominated by the ideas and staffs of the social work, education, and related social service and welfare professions, the

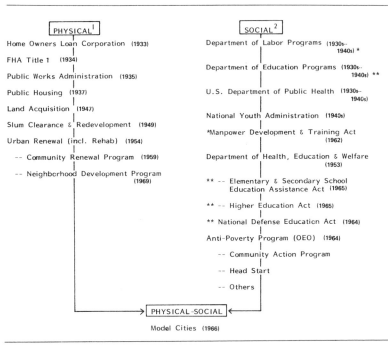

FIGURE 1.2   Parallel Development and Program Confluence of Physical and
                       Social Planning in the United States

1. SOURCES: Carter McFarland, "Residential Rehabilitation: An Overview," in
Carter McFarland and Walter K. Vivrett, eds., **Residential Rehabilitation** (Minne-
apolis: University School of Architecture, 1966), pp. 3-4; Charles Abrams, **The City
Is the Frontier** (New York: Harper & Row, 1965), pp. 241-242; and Langley Carleton
Keyes, Jr., **The Rehabilitation Planning Game** (Cambridge: MIT Press, 1969), p. 229.
2. SOURCE: **Report of the National Advisory Commission on Civil Disorders** (New
York: New York Times Company, 1968), pp. 141, 188-189.

pre-1964 federal housing, urban development, and related programs appear
to have been dominated by the *mystiques* and *motifs* of city planning,
management, engineering, and similar technocratic orientations. The 1964
antipoverty program, in a limited, informal way, and the 1966 Model Cities
program, officially, and to a greater extent, turned out to be mixtures of the
two separate pre-1964 orientations and staffings.

The Model Cities legislation was written explicitly for the integration and
coordinated execution of physical and social planning.[36] It aimed explicitly

at a locally worked out relative balance of the planning, designing, and rehabilitation (or redevelopment) of the built physical environment; *and* at planning, designing, or redesigning and development of social, economic, and cultural environments, processes, and structures. The key phrase summing up this general coalescent objective is "quality of life," that is, the total essential worth and value of individual and community life experience.[37]

Roughly 80 percent of the 117-acre Queensgate II project fell within the project boundaries of Cincinnati's Model Cities area. And although it included Appalachian whites, the bulk of the "model" area's population was black, as in many Model Cities areas across the nation.[38] For these reasons, there was from the beginning close coordination of Model Cities and urban renewal planning in Queensgate II. And by that token Queensgate II planning was more firmly related to the rest of the West End, to the Over the Rhine and Mt. Auburn neighborhoods.

The level of coordination and integration of physical and social planning in the Queensgate II project was enhanced by two lively factors: the citizen participation slant and social emphasis in the renewal planning, combined with the mandated physical and social role and purpose of Model Cities. The Queensgate II plan reflected these. Of the nine sets of policies, six were either explicitly or predominantly social in nature or equally social and physical, rather than predominantly physical: housing, family relocation, education, recreation, social service facilities, and job opportunities.

As in similar projects in Boston and other cities, then, Queensgate II in Cincinnati represented an interesting confluence of two major programs that in turn furthered the integration of urban physical and social planning. In some ways this marriage of "hard" and "soft" planning and development provided a compatible environment and grounding for the espousal of policy planning as a third major development in the planning field.

Policy planning is one of the newest looks in city planning. It began in the early sixties,[39] and its advent as a really potent idea occurred late in the same decade. It is the latest phase in what was a trend away from regarding land use, design, and technical and rational methodologies as the core staples of city planning. (Unlike advocacy planning, it is not narrowly identified with professional planners.) It seeks to broaden urban planning to more deliberately embrace social and economic factors and forces in its concepts and proposals. Herbert Gans was further outlining and publicly recording this new direction at the very start of the seventies.[40] Policy planning has arrived as an idea. It is not necessarily a widely practiced reality, but it appears to be established even among practitioners as a wave of the near future.

The sociohistorical procession of planning ideas from master planning, to comprehensive planning, to policy planning is, like advocacy planning, reaching to policy (see Figure 1.3). Policy may be defined as a guide to social action.[41] Insofar as "action" is construed to be active, overt behavior involved in actually building houses and highways, administering agencies and programs, and enforcing laws and regulations, then policy is the decision-making, law-establishing, and promulgation that guides that action.

Policy planning is a logical extension and expansion of Gans's well-known idea of about ten years earlier of goal-oriented planning. Goal-oriented planning regards the achievement of a community's goals, and associated appropriate, facilitating programs, as being far more important than becoming preoccupied with technical methods and procedures and professional techniques.[42] In the logic of policy planning, "goals" are the most definitive *"policy"* part of a plan, or of the planning process. Orientation to policy, and the implied goals involved, would force the planner, as in goal orientation, to "concentrate on the people and on the social and economic forces which foster their deprivation, rather than on neighborhood conditions which are themselves consequences of these forces."[43] Policy planning is goal-oriented planning *writ large*.[44]

Policy planning is the systematic formulation of social policy. In another place, Gans defines social policy as "any proposal for deliberate activity to affect the workings of society or any of its parts."[45] He sees goals, programs, and consequences as being the three major components of social policy design, or policy planning. In this sense, both sociologist-planner Gans and

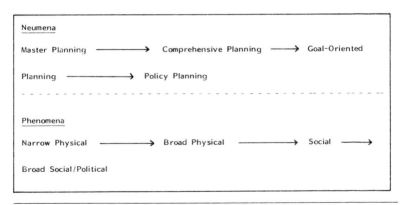

FIGURE 1.3   Changing Caste of Urban Planning

economist-planner Guy Benveniste see planning and social policy as virtually the same thing, differing only in scale.

Gans:

> As I see it, social policy differs from social planning only in scale; a plan is nothing more than a set of interrelated social policies.[46]

Benveniste:

> There is little difference between an announced set of policies and a published plan. . . . A plan is the outcome of a planning process. . . . Both policies and plans are statements of intents, and they are statements about the characteristics of future events.[47]

In this broadened sense, the city planner would be only one of a wide variety of experts engaging in policy planning.[48] And in this form, planning becomes consciously relevant to politics.

Associated with social policy and policy planning are policy analysis and policy research, all of these together regarded as constituent elements of the "policy sciences." Certain scholars, like Harold Laswell, cautioned against confusing the policy sciences with the social sciences.[49] Two other cautionary distinctions are sometimes mentioned and both are relevant to analysis or interpretation of Queensgate II planning; namely, policy research versus applied social research, and policy planning versus a policies plan.

Laswell once stressed that the policy sciences use social science *results,* insofar as those results have bearing on policy needs in a given situation. Gans complains, meanwhile, that sociology and other academic social sciences (presumably including much of the "applied" variety) are not policy-oriented, and therefore are not "suitable" to either policy or policy planning. But both policy research and some (nonpolicy) applied research are oriented to social action and are intended to influence it.

Policy research, however defined, obviously is an aid to policy planning and policy making. As Benveniste says, "Social policy research is one dimension of the process of policy-making."[50] But both logically and realistically, it is possible to formulate and draft a policies plan (that is, a plan which is policy in any case) without much conscious attention to the *planning of policy*. Likewise, it is logically possible to engage in policy *planning* (with or without policy *research)* for a long period without actually coming up with a policies *plan*.

The bottom line, however, is that in the field of urban planning, a policies plan is quite different in nature from the blueprint-prescription character of the old master plan.

Policy planning, then, amounts to the coordination of stated social, economic, political, and cultural goals and principles, for the purpose of guiding or orienting a society or a community toward a desired future state of affairs. This kind of planning is much broader than blueprinting the rearrangements of land use, redesigning neighborhoods and freeway systems, or engineering the redevelopment of cities. And policy planning is more likely to be interdisciplinary than conventional planning.[51] The main purpose of employing the University of Cincinnati in Queensgate II was to use academic, multidisciplinary, and interdisciplinary organized research in the development of an urban renewal plan.

Cincinnati's Queensgate II project consisted of a moderate level of policy planning, policy research, and applied social and urban research. This clearly marked the Queensgate II work as "new direction" planning. Applied research was conducted by (a) planners and architects, and (b) social scientists. The pros and cons of this interdisciplinary effort are discussed in Chapter 9.

The *policy* nature of the university team's planning was conditioned largely by the clear-cut policy groundwork and form of the West End Task Force's decision making.[52] The WETF policy decisions for the development of Queensgate II revolved around population (people) and physical development (mostly land use), pointed up in six policy "objectives":

(1) Resolve conflicts between residence and industry.
(2) Establish a character for the area west of Central Avenue.
(3) Establish a pattern for the area east of Central Avenue.
(4) Eliminate substandard housing.
(5) Minimize the negative social effects of clearance of the existing population.
(6) Improve circulation.[53]

The research and policy planning that took place in the Queensgate II project, 1967-1970, in any case, was indicative of the long-term tendency in the United States away from the primacy of land use planning and master planning.[54]

In summary: citizen participation, advocacy planning, integration of physical and social planning, and policy planning were hallmarks of the metamorphosis in urban community development that took place in the sixties, providing new directions for the seventies. The Queensgate II urban renewal planning program in Cincinnati may be seen as a microcosm of those changes in the United States.

NOTES

1. Richard L. Simpson, "Beyond Rational Bureaucracy: Changing Values and Social Integration in Post-Industrial Society," *Social Forces* 51, 1 (1972). Revision of presidential address. Sociological Society, 1972, reprinted by Warner Modular Publications.

2. Herbert J. Gans, "The Goal-Oriented Approach to Planning," paper presented in a joint meeting of the Puerto Rico Planning Society and the Puerto Rico Economic Association, San Juan, 1958.

3. Herbert J. Gans, *People and Plans* (New York: Basic Books, 1968), 78-83. In an introduction to a group of his essays on goal orientation, Gans writes, "If the physical environment plays only a minor role in people's lives and the community is not the most important social group in which these lives take place, traditional planning comes close to being irrelevant" (p. 53). Goal orientation is an alternative.

4. Alan A. Altshuler, *The City Planning Process: A Political Analysis* (Ithaca, NY: Cornell University Press, 1965). In a strongly stated paragraph, he concludes: "When government must act to deal with some pressing issue, every effort should be made to define the problem narrowly and to deal with it specifically. The approach should be one of dealing with bottlenecks, not planning the whole production line. In other words, it should be piecemeal, not comprehensive" (p. 313).

5. Altshuler, *City Planning Process*.

6. For example, Miami-Dade County in Florida and Indianapolis's Marion County "Univgov" in Indiana.

7. Mel Scott, *American City Planning Since 1890* (Berkeley: University of California Press, 1969), pp. 85, 210, 397, 423, 431-33, 447, and 495 for this and immediately following information.

8. This apparently incorporated many of Jacob Riis's ideas and anticipated some of Jane Addams's humanistic fervor.

9. Noel P. Gist and L. A. Halbert, *Urban Society,* 4th ed. (New York: Thomas Y. Crowell, 1956), pp. 482-483.

10. Arthur B. Gallion and Simon Eisner, *The Urban Pattern* (New York: D. van Nostrand, 1963), p. 320.

11. Gans, *People and Plans,* p. 71.

12. Even planning commissions, composed of high-ranking citizens, have had no power to put their plans into operation. Gallion and Eisner, *The Urban Pattern,* p. 482.

13. For one explanation as to why, see Gans, *People and Plans,* pp. 70-71 and 242.

14. The urban riots represented the extreme forms of social action aimed at economic fair-share, equal justice, and resolution of grievances. But these were preceded by milder forms of social unrest, civil rights demonstration, and urban protest, stretching from 1963 to 1966. See Introduction to *Report of the National Advisory Commission on Civil Disorders* (New

York: Bantam Books, 1968), pp. 35-40.

15. *Report of the National Advisory Commission on Civil Disorders,* mainly pp. 47-52 and the third series of charts attached at the end of the report. Also, of the 23 studied, the 10 "profile" cities: Tampa, Cincinnati, Atlanta, Detroit, and the New Jersey cities of Newark, Plainfield, New Brunswick, Elizabeth, Englewood, and Jersey City (pp. 35, xxi, and 69).

16. Charles Abrams, *The Language of Cities: A Glossary of Terms* (New York: Viking, 1971), p. 61.

17. Abrams, *The Language of Cities,* p. 61.

18. Cincinnati *Enquirer,* January 30, 1969: "New Krabach Proposal: Make 'Seven City Halls.'" Richard Daley in Chicago, Kevin White in Boston, and mayors in other cities had started the "little" city halls movement earlier.

19. George Frederickson, ed., *Neighborhood Control in the 1970s: Politics, Administration, and Citizen Participation* (New York: Chandler, 1973), p. viii.

20. The comparatively few dramatic efforts at "community control" in white communities have been around such issues as opposition to busing for school integration, fighting to set up special community schools, resisting annexation by larger communities, and resisting ties to metropolitan transit systems that have financial problems. Many will explain that few white communities are involved in the community control "movement" simply because, by comparison with most black communities, whites *are in control* of their communities.

21. Alan A. Altshuler, *Community Control: The Black Demand for Participation in Large American Cities* (New York: Bobbs-Merrill, 1970).

22. Peter H. Rossi and Robert A. Dentler, *The Politics of Urban Renewal: The Chicago Findings* (New York: Free Press, 1961). See especially Chapter 5.

23. ARCH began as a lily-white group of advocate architect-planners; but after rumbles in the Harlem community, blacks were added until it became virtually an all-black staff. I was one of the first blacks asked to work there, but it was at a time I could not leave my Boston urban renewal position.

24. Robert Goodman, *After the Planners* (New York: Simon and Shuster, 1971), pp. 20ff.

25. Langley C. Keyes and Edward Teitcher, "Limitations of Advocacy Planning: A View from the Establishment," *Journal of the American Institute of Planners* 36, 4 (July 1970): 225-226.

26. John Friedmann, *Retracking America: A Theory of Transactive Planning* (Garden City, NY: Doubleday 1973), pp. 279-281.

27. See Saul Alinsky, *Reveille for Radicals* (Chicago: University of Chicago Press, 1946).

28. Joan Levin Ecklein and Armand A. Lauffer, *Community Organizers and Social Planners* (New York: John Wiley and Sons, and Council on Social Work Education, 1972), p. 1.

29. Paul Davidoff, "Advocacy and Pluralism in Planning," *Journal of the American Institute of Planners* 31, 4 (November 1965). See also Goodman, *After the Planners,* pp. 171-174; and Gans, *People and Plans,* p. 73.

30. Paul Davidoff, "Advocacy and Pluralism in Planning," in H. W. Eldridge, ed., *Taming Megalopolis,* Vol. II (New York: Anchor, 1967), p. 602. See also Susan S. Fainstein and Norman I. Fainstein, "City Planning and Political Values," *Urban Affairs Quarterly* 6, 3 (March 1971): 345ff.

31. Too voluminous to list exhaustively, it includes, since 1965, articles by Peattie, Keyes and Teitcher, Mazziotti, Gilbert and Eaton, the Davidoffs and Gold, Kaplan, and others, in the *Journal of the American Institute of Planners;* articles in other periodicals, such as *Planning*

(especially 1968 and 1971), *Ekistics,* and *Urban Affairs Quarterly;* an editorial in *Progressive Architecture;* a symposium in *Social Policy;* and a book by Hatch. Reviews of and references to these writings in periodicals and books, respectively, have also increased.

32. Hatch is the creator of the "urban action" concept and program through which many architects and planners have entered the advocacy field. ARCH was an early example. Goodman is the author of the book *After The Planners* (1971), which still is influencing contemporary young professionals and graduate students, as future professionals. For a European example of an advocate *architect*-planner, see Thomas H. Jenkins, "Dolci: Passionate Planner for People," *Ekistics* 27, 158 (January 1969): 67-69.

33. Volunteers in Service to America, one of the programs within the Department of Health, Education, and Welfare, which operated actively during the Johnson Administration and was diminished and diluted during the Nixon years.

34. About the same time of the VISTA, CDC, and AIA-VISTA projects, some planners were carrying out advocacy work in such instances as the continuing work of UPA (MIT's Bernard Frieden, Harvard's William Nash, et al.) and Harvard's Urban Field Service (UFS) neighborhood assistance program, inspired and conducted by Chester Hartman. Davidoff's advocacy Suburban Action program (complementing Hatch's Urban Action) came later, at the turn of the decade.

35. Although the Ford Foundation had set in motion a number of "human renewal" projects in 1961, by 1964 so-called human renewal had not been effectively linked to the ten-year-old federal Urban Renewal program, even on paper. See Gans, *People and Plans,* pp. 204-205. For evidence and an illustration of coordinated physical and social planning in action, see Robert J. Perlman, "Social Welfare Planning and Physical Planning," *Journal of the American Institute of Planners* 32, 4 (July 1968): 237-241.

36. Gans, *People and Plans,* p. 205. On this general subject, see also: Harvey S. Perloff, "Common Goals and the Linking of Physical and Social Planning," in Bernard J. Frieden and Robert Morris, eds., *Urban Planning and Social Policy* (New York: Basic Books, 1968), pp. 346-359; and Gans, "Social and Physical Planning for the Elimination of Urban Poverty," in *People and Plans,* pp. 39-54.

37. In addition to Model Cities and the antipoverty programs, a parallel effort to effectively interrelate physical and social schemes (that is, environmental design) for human betterment is to be found in the burgeoning field of behavior and design (B & D). The international Architectural Psychology Conference, the mostly American Environmental Design Research Association, the British-based Environmental Psychology group, and a plethora of B & D publications and research centers provide lively evidence of this current movement.

38. Edward J. Logue, while Development Administrator for Boston, used to refer to the Model Cities legislation as "the Ghetto Bill."

39. See particularly, Leonard J. Duhl, ed., *The Urban Condition: People and Policy in Metropolis* (New York: Basic Books, 1963). One of the contributions was the signal essay by Melvin Webber, "The Prospects for Policies Planning," pp. 319-339.

40. Herbert J. Gans, "From Urbanism to Policy Planning," *Journal of the American Institute of Planners* 36 (July 1970): 223-225.

41. James S. Coleman, *Policy Research in the Social Sciences* (Monograph), (Morristown, NJ: General Learning, 1972), p. 2.

42. See Gans's short essay, "The Goal-Oriented Approach to Planning," in *People and Plans,* especially pp. 79-80, 78-83; also p. 53.

43. Gans, *People and Plans*, p. 240. Interestingly enough, Ronald M. Caines pointed to the emerging natural relationship with Advocacy Planning: "Advocacy for the Seventies Will Aim at Public Policies," *AIP Newsletter* (January 1970). Also, about six months later, Frances Fox Piven's advocacy article, "Disruption is Still the Decisive Way," appeared in *Social Policy* 1 (July-August 1970): 40-41, preceded by a "Symposium on Advocacy Planning," earlier in the same journal (May-June 1970): 32-37.

44. Guy Benveniste, for example, says that policy experts and planners "[define] expert knowledge as the knowledge required to link desirable outputs (i.e., values and goals as determined in the political process) with needed inputs (i.e., specific government programs). See Benveniste, *The Politics of Expertise* (Berkeley: Glendessary Press, 1972), p. 64.

45. Herbert J. Gans, "Social Science for Social Policy," in Irving L. Horowitz, ed., *The Use and Abuse of Social Science: Behavioral Science and National Policy-Making* (New Brunswick, NJ: Trans-action, 1971), p. 14.

46. Gans, "Social Science for Social Policy," p. 14.

47. Benveniste, *The Politics of Expertise*, p. 34.

48. For example, the group of professionals (including Gans) who, after informal meetings and considerable dialogue, produced the papers in the Duhl book on policy in relation to people and the urban environment, were: "psychoanalysts, public health physicians, psychologists, animal ecologists, sociologists, biologists, city planners, journalists, humanists, and scientists" (Duhl, *The Urban Condition*, p. xiii).

49. See Benveniste, *The Politics of Expertise*, p. 63.

50. I take it that a corollary to Gans's criticism of non-policy-oriented social science is that social planners and city planners busily and dutifully produce plans that are *in effect* social policy, but that many of their plans are not the result of a rational coordination of basic policies and appropriate accompanying programs. Such plans are therefore less fundamental and less effective.

51. Benveniste, *The Politics of Expertise*, p. 6.

52. West End Task Force, *Phase II Policy Decisions* (December 1967), p. 16.

53. West End Task Force, *Phase II Policy Decisions*, p. 16-21.

54. For a similar but differing interpretation of these developments, see Harvey Perloff, "The Evolution of Planning Education," in David R. Godshalk, ed., *Planning in America: Learning from Turbulence* (Washington, DC: American Institute of Planners, 1974), pp. 161-180. For *contrary* evidence, see Jerome Kaufman, "Contemporary Planning Practice: State of the Art," in the same volume, especially pp. 119-129.

CHAPTER 2

Queensgate II:

A HISTORY OF A NEIGHBORHOOD

Zane L. Miller

It is difficult to write the history of a neighborhood in an American city because the term, historically applied, connotes nothing more than a vaguely defined geographic area. This is due in large part to the fact that community, in a social-geographic sense, has been a scarce, virtually nonexistent phenomenon in American urban life. Since the early nineteenth century, most great American cities have been growing and volatile entities, and their dynamism has guaranteed intramural shifts of population and local functions. As a result the history of American neighborhoods has been a history of chronic diversity, instability, and movement. Community, in this context, could not develop as an adjunct of a specific geographic locale out of the natural course of events. It is not too much to say that change and discord, not cooperation and community, have been the salient characteristics of neighborhood life in urban America.

This does not mean, however, that a sense of community has not developed within American neighborhoods. Quite the contrary. But it has not grown out of similarity of attitudes and goals, homogeneity of populations, or congruity between physical form and socioeconomic structure. Rather it has been more often contrived and based upon the tensions, anxieties, and conflicts generated by ceaseless technological, population, and physical changes both within and beyond the boundaries of any given neighborhood. Consequently both the sense and particular manifestation of community in a specific area have been ephemeral events. The history of a neighborhood, in short, is bound up with the history of the city, and its main theme is not continuity but flux.

From Suburb to Urban Periphery, 1830-1850

Queensgate II, of course, until very recently, was not even a neighborhood, much less a community. In 1830, eleven years after it became a part of Cincinnati, it was a suburb not uncharacteristic of similar regions on the fringes of other American cities. Cincinnati's central business and commercial district then focused on Main Street and the east-west streets in the Ohio River bottoms. To either side and above this area on Third Street the city's affluent minority had established a mixed but respectable residential district. Around this core lived the middling orders of society. Beyond them the poor, the most recently arrived native newcomers (black and white), and the greenest foreign immigrants occupied the more tightly built-up portions of the urban periphery. Still farther out, though partly within the city limits, lay a semibucolic and semiindustrial suburban fringe.

Queensgate II in the 1830s fell into the last category. It was part of a large region above Fourth and beyond Western Row (Central Avenue) known as the West End. Though literally on the edge of town, it was an important part of the city. From the Betts house (now 416 Clark Street)[1] north stretched a broad and grassy though irregular plain known to contemporaries as "Texas." Occupied by dairies, farms, and truck gardens, one of its chief functions was to feed the city. As late as 1838 most of the 250 acres of pasturage within the municipal boundaries lay in that region.

By that time, however, the land was already being put to other uses. After 1815, as meat packing became an increasingly important segment of the city's economy, slaughterhouses, soap factories, and tanneries began to appear in Texas. Closer to the urban periphery, on what is now Cutter Street, near Court, stood Methodist, Baptist, and Roman Catholic cemeteries. And to the east, in a plot of land on Chestnut near Central, was the burial ground of the city's English Jewish residents, perhaps the first Jewish cemetery east of the Alleghenies.[2] None of these denominations ranked among the city's socioeconomic elite and the presence of their cemeteries in Queensgate II reflected the lowly status of the location.

The combination of market farming, meat packing, and cemeteries, for a variety of reasons, retarded suburban residential development. In the first place, these facilities would have to be bought up and moved before dwellings could go up. Second, the meat packing and allied manufacturers generated offensive odors. But worse still, these industries and the cemeteries, according to prevailing scientific opinion, emitted poisonous, disease-bearing gases, and the swamp land in the Mill Creek bottoms exacerbated the same problem (air pollution is one of the few characteristics of the West End which has persisted throughout most of its history as part of Cincinnati).

Despite these obstacles to residential development, a real estate boom began in the 1830s which altered the entire structure of the West End from the Ohio River to Texas and carried the region into the next phase of its history. The impact of the boom was particularly heavy in Queensgate II. For a brief moment the suburb of the 1830s seemed destined to become, like other suburbs before it, a peripheral slum.

Several factors triggered the boom. Between 1830 and 1850 the city's population rose from 26,515 to 115,435. This, along with the new street and building construction which accompanied it, placed a high premium on land within the developed portions of town. But walking city technology simply could not accommodate the new demand. Buildings could not be made taller and, since people still had to walk or ride in horse-drawn vehicles to get about the city, extensive outward expansion was out of the question. Movement to the east, moreover, was blocked by the hills. The crush for space set off by these circumstances helped to create an explosive situation.

Conditions in the lower West End were particularly acute. There the anticipated completion of the Whitewater Canal attracted commercial establishments at about the same time that the gas works was built. Simultaneously the central business district was pushing relentlessly to the west.[3] These developments squeezed Little Bucktown—a Negro district in the Ohio River bottom section of the West End also known as the Swamp—the heavily Irish district to the west and north, and the aristocratic section of west Third between the jaws of an ever-tightening vice and contributed to racial tensions which, in 1836, erupted in a race riot in the Swamp region. The same pressures also set the stage for the land boom which by the 1850s had made the Queensgate II section of the West End one of the most densely settled parts of the city.

There were only two areas open for the housing of new people, for business expansion, and for the resettlement of the displaced. One was north along and above the Miami and Erie Canal. Manufacturers and merchants were attracted to canal sites by transportation advantages and relatively cheap land. Immigrants, principally Germans, because of the comparatively low cost of housing, were also lured across the canal and created an area dubbed Over the Rhine.

The other safety valve was the Queensgate II region and Texas. Along west Fourth and Fifth modest frame buildings antedating the 1830s were demolished and replaced by brick dwellings as the Third Street elite neighborhood expanded. To the north, from Sixth to Laurel, a diverse conglomeration of lower- and middle-class Cincinnatians overran the farms, factories, and cemeteries and established their homes and institutions.[4]

Census takers and Charles Cist, the town's chief booster, caught the outlines of the development. In 1840 the Seventh Ward encompassed all the West End. Its population was 4740, roughly one-third (1473) of whom were Germans, and but 73 Negroes. More importantly, the inhabitants were almost evenly divided between males and females. This was a sure sign that it was not being occupied solely by newcomers, for recent inmigrants characteristically were made up principally of males, who, preceding their families or too young to have families, tested the opportunities of the city before settling permanently.[5] Most of the new construction, moreover, was of frame rather than the more permanent and expensive brick, although by 1845 the proportion of brick to frame buildings seemed to be evening up.[6]

A survey of the institutions within the area reinforces the conclusion that the population was diverse. By the mid-nineteenth century Queensgate II contained 22 nonbusiness institutions. One, the Pest House, established in 1820 at Twelfth Street on the Miami Canal (Central Parkway), was a leftover from the area's pre-1830 suburban period. Neither it nor the orphanage on Kemble between Plum and Elm was the kind of institution associated with fashionable or even middle-class districts. Of the region's four suburban cemeteries, only the Jewish on Chestnut had survived the post-1830 boom, and 1849 marked the last interment there. The majority of institutions in the area at midcentury, then, appeared during the boom.

The most prevalent single type of institution was religious. There were seven churches located within Queensgate II. Only one, that of a Protestant Episcopal congregation at Seventh and Plum in the southeast corner of the region, ranked among the city's socially prestigious religious groups. There was also a German Catholic church (St. Joseph's) at Linn and Laurel, and a German Reform building immediately to the north of Queensgate II on Betts between John and Cutter. St. Peter's Cathedral, principally an Irish institution in these years, stood at Eighth and Plum. Only one Jewish group, the Polish Congregation of the Community of Israel, had a building within Queensgate II proper (at Sixth and Central), but another on Lodge between Fifth and Sixth (Holy Congregation of United Brethren) served the increasing number of German Jewish immigrants moving into the region.

There is a paucity of evidence about other structures in the region at midcentury. There were but two fire companies in the entire West End above Sixth (one on Sixth between Central and Plum and the other on Cutter above Clark), and the only water available for fire fighting had to be drawn from cisterns along Central Avenue. The area's two colleges, a medical and a commercial school, were both situated on the southern fringe of Queensgate II close to the expanding aristocratic West End along Fourth, Fifth, and Sixth. The sole public school sat in the same area, near John. The region, moreover, offered few job opportunities. It contained, according to one survey, no banks or hotels, but three factories, merely a handful of commercial establishments, and even fewer professional offices.[7] Most residents, apparently, had to walk considerable distances to work or to obtain essential services. The documents, in short, contain nothing to contradict the impression that at midcentury, the region was, at best, a marginal neighborhood housing a low- to middle-income and racially, ethnically, and religiously mixed population.

Periphery to Center, 1850-1880

Queensgate II's tenure as a peripheral urban district was brief. From 1850 to 1880 the city's population grew from 115,453 to 225,139, despite its relative lag in the race for urban supremacy in the trans-Appalachian West. More significantly, the physical area of the city in the same time span expanded from 6.7 to just over 22 square miles. Nonetheless, as late as 1875 the overwhelming bulk of Cincinnati's population, rich and poor, still lived within the Basin and the lower Mill Creek Valley. Consequently the distension of the municipal boundaries scrambled but did not obliterate the historic pattern of the walking city. But it did produce shifts in the intramural composition of neighborhoods. In the process, Queensgate II was transformed once again.

The change was brought about by a combination of old and new forces. The most important new element in the situation was the introduction of the horse-drawn street railway. The city's first line began in 1859, and within twenty years there were four lines operating in the West End. The cars, running regularly across Sixth, Seventh, Eighth, and Mound, and north-south on Central, John, Linn, and Freeman, made every portion of Queensgate unprecedentedly accessible. Perhaps the most important single line was that extending from Brighton—the city's newest industrial suburb—down Western Avenue and Baymiller, across Fifth and Sixth through the central business district into the Eggleston Avenue factory district, and back through the fashionable East End around Third and Pike Streets.[8]

At the same time older forces continued to uproot individuals and private and public institutions. The lower West End below Fourth from Central Avenue to Freeman continued to fill with railroad tracks, terminals, and warehouses as the city established long distance rapid transit ties to the Southwest and Chicago. More residential space was absorbed as hotels, most notably the Burnet House at Third and Vine, the Grand at Central and Fourth, and the Carlisle at Sixth and Mound, occupied the strategic ground between the West End railroad facilities and the central waterfront and business district.

Simultaneously the central business district itself was undergoing a marked expansion and reorientation.[9] Third Street emerged as the prime locus of the city's financial institutions, and Vine Street and Central Avenue between Third and Sixth became the heart of the metropolitan central business and retail district. These developments not only all but obliterated Little Bucktown but also unmoored the Irish neighborhoods around it and displaced the prosperous families that had settled West Third and Fourth. Finally, the reshuffling also helped develop an industrial-residential region on the northwestern periphery of the city which extended along the railroads in Mill Creek Valley as far north as Cumminsville.

This complex process drastically altered the situation of Queensgate II. It now rested not on the urban periphery, but adjacent to the center of the city. Yet neither it nor the contiguous parts of the West End constituted a slum. It was, in fact, during this period (1850-1880) that Queensgate II and its vicinity earned a reputation as one of the most respectable and substantial residential districts in the city. Although the scope of this study does not provide for a systematic house-by-house check, the evidence suggests that the legend is reasonably accurate. Architectural vestiges, guide books of the period, land use maps especially prepared for this project, and a variety of other recent studies indicate that at one time or another between 1850 and 1880 virtually every street within the present boundaries of Queensgate II housed at least some of the silk-stocking crowd. But the elite never formed a solid enclave, and their residences constantly drifted north. Dayton Street, in fact, was their last, brief stop before they evacuated the West End entirely. The neighborhood, to be more specific, was just as "marginal" in its heyday of affluence as it had been when it lay on the urban periphery.

Despite the prosperity of the 1850-1880 years, Queensgate was not a WASP heaven. If the district had a predominating ethnic flavor, it was German Jewish. This, however, was the second, not the first, site of Jewish settlement in Cincinnati. Before the mid-nineteenth century, both English and German Jews had congregated in the East End. By 1852 the tide of German Jews to Cincinnati from Bavaria and other south German states which began in the 1830s had overwhelmed the English congregation and resulted in the construction of a new synagogue at Sixth and Broadway. This, however, occurred at just the time that many East End Jews, like other whites, were fleeing east Sixth Street before the expansion of Bucktown, a black ghetto—the city's third[10]—which developed in the Deer Creek bottoms industrial zone after the completion of the Miami Canal. The old Broadway Temple was sold to an African Methodist Episcopal group and, in 1866, Dr. Isaac M. Wise's Reform congregation moved into its new and architecturally distinctive synagogue at Eighth and Plum across from St. Peter's Cathedral and one block down from City Hall. Three years later another German Jewish group occupied its new synagogue at Eighth and Mound in a building only slightly less opulent than that at Eighth and Plum. And before 1880 yet another Jewish congregation, also German, was worshipping in a synagogue on Lodge between Sixth and Seventh.[11]

Other Jewish organizations also met in Queensgate II. Although Reform Jews moved very freely in the upper strata of Cincinnati society, the process of acculturation was not yet complete. Consequently, as the club mania (a phenomenon not unrelated to the physical expansion of the city) gripped the city's upper classes, the German Jews also established elaborately equipped facilities for social and recreational purposes close to the bulk of their population. One, the Allemania Club (established in 1849), moved from Main to Walnut in 1863 and arrived at its "elegant" location at Fourth and Central in 1879. Equally if not more exclusive was the Phoenix Club, a social and literary society formed for the express purpose of preserving the bonds of the community.[12] Hebrew Union College, a theological school established in 1875 at Sixth Street west of Cutter, served much the same function.

The other Jewish institutions in the area grew out of continued German Jewish immigration and sought out the less successful newcomers. Throughout the 1870s the Hebrew Relief Association met at Covenant Hall at the southwest corner of Fifth and Central Avenue, and six of the seven B'nai B'rith mutual benefit lodges met across the street at the northwest corner of Fifth and Central Avenue. Together these groups provided sick benefits, life insurance, and cash and supplies to sick, poor, or transient Jews.[13] Like the other Jewish institutions, then, these tended to cluster in the southeast corner of Queensgate II near St. Peter's and the Plum Street Temple.

Yet this agglomeration of Jewish and Catholic institutions was not a peculiar feature of the neighborhood. In fact, these institutions sat in the midst of several Protestant churches, creating a holy concentration which led one contemporary to dub the district "the Church Region."[14] By far the wealthiest of these congregations, judging by the scale of the buildings, were the St. Paul's Methodist Episcopal and the Second Presbyterian. The former, completed in 1870 at a cost of a quarter million dollars, was at Seventh and Smith Street, and the latter, completed in 1875, was at Eighth and Elm. Both, moreover, had moved up and over to the Queensgate II area from sites on Fourth Street.[15] Other substantial Protestant churches to enter the district between 1850 and 1880 were the Congregational Unitarian Church at Eighth and Plum (1869-1870), and the Central Christian Church on Ninth between Plum and Central Avenue (1869). The latter had had a particularly difficult time maintaining its centrality in the face of the constant urban flux of the nineteenth century. Its original location had been on Sycamore above Fifth, and in 1845 it moved from there to Eighth and Walnut before settling in Queensgate.[16] The presence of these structures, plus the establishment in 1870 of the Queen City Club (a haven for the city's most successful businessmen) at Seventh and Elm, within walking distance of the homes of most of its members,[17] strongly suggests that affluent Christian as well as prosperous German Jews lived in and around Queensgate II as late as the 1870s.

By the time the Church Region was built, however, Queensgate II was a split-level community. Generally speaking, Ninth Street was the dividing line, and the land below it was devoted principally to nonresidential uses.

The density of religious structures, of course, cut sharply into the land available for dwellings. Robinson's Opera House took up another chunk at the southeast corner of Ninth and Plum, and in 1853 City Hall was erected at Eighth and Plum, its present site. By 1875, moreover, because of its contiguity to the central business district, a wealthy residential community, and City Hall, the plot encompassed by Race, Sixth, Central, and Ninth contained more doctors' and lawyers' offices than any other similar sized section of the city. And Eighth Street around Garfield Place was being encroached upon by the expanding downtown club district.[18]

Nor was that all. Queensgate II below Court Street straddled Central Avenue, a major north-south artery between the Miami Canal and the Ohio River, and it was virtually adjacent to the lower West End railroad complex. It was, moreover, close to retail outlets in the central business district, and its street railway system made it accessible to workers all over the Basin. As the available factory sites in the Ohio River bottoms and lower West End shrank, manufacturers began to move into the section to exploit its potential. By 1875 there were 23 industries in Queensgate II, all of them below Court between John and Elm Streets. Some, such as the 4 carriage and furniture manufacturers, though fire hazards, were relatively "clean." Others, such as the 13 metal works and machinery shops, were noisy. The others—3 soap, chemical, and candle producers, 2 brewers, and 1 meat packer—were both noisy and smelly. Together these factories formed a half-circle around City Hall, the Church Region, and the professional district.

Commercial facilities filled in the nonresidential interstices around the heart of Queensgate II. In 1875, the area contained 87 retail shops. They were not, however, randomly scattered, but clustered in 5 groups: 1 lay to the south in the 2 blocks between Plum and Central and Sixth and George; another focused on the intersection of Mound and Ninth; the third centered around the block bounded by John and Central and Ninth and Eighth; and the last occupied most of the block between Richmond and Court and Central and Plum. Since many shopkeepers lived above or next to their stores, however, these plots served both a commercial and a residential function.

To the north and northwest of the City Hall, church, factory, and commercial core as far as Cutter on the west and Laurel on the north, the region was largely residential, and largely, as indicated above, middle- to upper-class. But not exclusively so. A random sample of residents by occupation in 1875 reveals a remarkable socioeconomic mix. Artisans, mechanics, and lower-level white-collar workers predominated, and many of them lived cheek to jowl with the Gambles, Doepkes, Rollmans, Seasongoods, and other representatives of the city's elite. Both the occupational survey and the 1870 census statistics, moreover, indicate that older first-generation and second-generation Irish and Germans, augmented by a sizable number of recent arrivals, constituted the bulk of the population. And the concentration of German and Irish Catholic churches in the immediate vicinity of Queensgate II indicates that the bulk of Germans, as well as the Irish, were Catholic.

From 1850 to 1880, then, Queensgate II constituted a diverse though not patternless neighborhood. Its boundaries placed it in two of the most populous wards of the city, districts which contained every kind of housing, from tenements serving six and more families, to second-story apartments, to town-house mansions. Both wards were close to the city average with respect to people per dwelling, and mortality rates also hovered near the city average. Together, the two wards housed 821 blacks, and two Negro Baptist churches had appeared on West Ninth, one at Mound and the other between John and Central. A Colored Orphan Asylum also stood on Ninth near Elm. But Queensgate II was becoming increasingly crowded and its streets were heavily traveled, for the heavy concentration of churches, factories, and commercial establishments on its southeastern border served as a funnel, its services pulling people to it from the expanding West End beyond.[19] The region was, in fact, what it had been in the 1840s, a marginal neighborhood. And it stood on the edge of a revolutionary development in the process of urbanization which obliterated the walking city and placed Queensgate II in the center of the city's most notorious slum. The institutional, industrial, and commercial boom in the lower parts of the neighborhood in the 1860s and 1870s were signals announcing its imminent deterioration.

The Lively Slum, 1880-1930

Close observers of the urban scene in 1874 were convinced that the city faced a crisis. To them it seemed that the Basin was doomed as a residential site, that it would soon be filled with "business and manufacturers." The "crowding," wrote one critic,

> increases crime. The morals become depraved, intemperance reigns, and men . . . become mentally and morally unfit to [work]. . . . Living in such a state of depravity, they become the ready tools of unprincipled and avaricious politi-

cians . . . who sink their constituents into a still deeper state of degradation;
hence our exhorbitant [*sic*] taxes, our heavy city debt.

It seemed a grim prospect. The only solution, he felt, was for city government or private capitalists to open up means of rapid and cheap travel
through or over the hills. Urban sprawl, he felt, held the key to a golden age
of pure air, pure morals, and pure politics for Cincinnatians.[20]

By 1900, the millenium had arrived. The construction of incline planes in
the 1870s and the rapid development of electric rapid transit after the mid-
1880s—by 1912 the city had 222 miles of tracks—provided the means for
the invasion of the hills. Continued immigration and the growth of manufacturing, coupled with the improved transportation technology, provided the
impetus which drove both factories and people out of the Basin. Cincinnati
burst out of its walking city seams. Its physical size extended from 22.20
square miles in 1880 to 35.27 in 1900 and over 50 in 1920. The Mill Creek
industrial belt, stretching from Mohawk-Brighton northeast to Oakley, was
clearly visible by 1905, and only a series of annexations prevented the
successful secession of the thousands of residents who overran the hilltops.[21]

The explosion created a new city. On the outskirts, inhabited by the
occupationally, ethnically, and religiously mixed but prosperous elite, lay
the suburbs. Closer in, between the core and the suburbs, lived a heterogeneous mass of blue- and white-collar workers in a region known to social
workers as the Zone of Emergence. Closer still, between the Zone and the
central business district, a vast circular slum took shape and began to press
relentlessly outward.[22] The region began in the Deer Creek bottoms,
stretched southward under the brow of Mt. Adams, curved along the riverfront, pushed up Mill Creek Valley through the West End and Queensgate II
to Mohawk-Brighton, and cut across the Over the Rhine district back to the
Deer Creek bottoms.

For Queensgate II the transformation into a new situation and status
began in the 1880s. The larger industries were among the first to vacate. In
the mid-1880s Proctor and Gamble abandoned its old plant at 830 and 832
Central Avenue and moved to Ivorydale, in the Mill Creek industrial belt. By
1900 only three substantial factories remained. Modestly prosperous commercial establishments evacuated next. In 1911, one realtor, concerned
about sagging real estate values in the region, estimated that in the last
decade the number of small stores in the West End had decreased by 50
percent. Although Central Avenue continued to function as a major commer-

cial street, it had, as early as the mid-1880s, become tawdry. From Fifth to
Mohawk, recalled one old resident, it was lined with shops, "small one man
affairs, operated by the owner, with perhaps one assistant."

> A furniture store at Fifth and Central got the parade off to a good start and was
> followed by pawn shops, a book store, groceries, barber shops, shoe stores,
> dry goods stores, cigar stores, and a host of others, plentifully seasoned with
> saloons.[23]

In the 1880s, moreover, the elite began to abandon the Queensgate II
region. By 1884, physicians, dentists, and residents of Garfield Place were
disgusted with the "disreputable and worthless vagabonds of both sexes"
who crowded the benches of Garfield Park. In 1892, a $25,000 home on
West Sixth sold for $14,000, and residents were complaining that the grow-
ing "cosmopolitanism" of the neighborhood was ending their "seclusion."[24]
By the turn of the century the process was virtually complete. "The fact is
historic," asserted one uprooted hilltop resident in 1899, "that when West
Sixth and Seventh and Eighth and Dayton Streets, . . . once filled up by the
homes of the elite . . . gentiles of Cincinnati, were encroached upon by
Hebrew families the Christians fled to Walnut Hills as the haven of rest." But
"when rapid transit came the Hebrews flocked to 'the hill' until it was known
by the name of New Jerusalem." Then "Avondale was . . . heralded as the
suburb of deliverance, but again rapid transit brought the wealthy Hebrews
to Avondale in numbers greater than the flock of crows that every morning
and evening darken her skies." And now, he concluded "it has been face-
tiously said that the congregation has assembled in force and that when
Avondale is roofed over the synagogue will be complete."[25]

Those who held on in Queensgate II lived in fear. Thomas Stanley Matthews, whose father was a priest at St. Paul's Episcopal Church (on Seventh
at Smith) after the turn of the century, was left with vivid memories. "The
door was not only latched but bolted and had a chain, always fastened at
night" as an added precaution against "sneak thieves." The "nightly battering
down of the house," he recalled, "increased our sense of siege." Thereafter
the "city," for him, though it had its virtues, was in the final analysis "a place
of horrors, an urban desert."[26]

One cause of this fear was a dramatic alteration in the composition of the
population. Overrun by newcomers, the region's traditional diversity was
intensified. From 1880 to 1910 immigrants from southern and eastern Eu-
rope led the invasion. A band of Genoans and a group of Levantine Jews

established separate but contiguous colonies along the southernmost fringe of Queensgate II along Fifth and Sixth. A Russian Jewish colony developed in and around Queensgate II, and Rumanian and Austro-Hungarian Jews settled to the north and spread up into Mohawk-Brighton.[27] Contemporary observers and census statistics suggest, moreover, that large numbers of first- and second-generation Irish and Germans, as well as native Americans, remained in the region. Nonetheless, the three wards into which Queensgate II fell in 1910 (Sixth, Fifteenth, and Eighteenth) contained a larger proportion of southern and eastern Europeans than any in the city except one (the Eighth, comprising the southeastern portion of the Basin). And the expanding black ghetto south and west of Queensgate II had, by 1900, begun to edge up across Longworth and George toward Seventh as far east as Plum.

Housing conditions in and around Queensgate II deteriorated rapidly. The absence of sanitary facilities and overcrowding were the chief problems. The Fifteenth Ward, which included the northern part of Queensgate II (Richmond and above), had almost 130 people per acre in 1913, the highest density in town. Homeownership in the same district was 8 percentage points below the city average, and only just over 2 percent of the Fifteenth's homes were encumbered by debt, an indication that few residents were able or willing to begin the process of buying a home. Not surprisingly, the 3 West End wards were among the city's leaders in mortality and cases of consumption.[28]

As a result, people went into the streets for relief from the congestion and tedium of close living. Yet they found little recreation. Like the rest of the slums, the area in and around Queensgate II was plagued with petty criminals, derelicts, and juvenile delinquents. Policy operators and bookmakers especially flourished. Worse still, the region also contained the city's most notorious red-light district. "On both sides of Longworth and George Streets," announced the Reverend John Howard Melish in 1903, "nearly every house is used openly for immoral purposes, with the madams' names upon the doors."[29] The West End tenderloin attracted the prosperous as well as the poor. It was here, for example, that one immigrant musician played for "an all night carousal" in one of the "aristocratic bawdy houses" which "some big butter-and-egg men put on for their visiting customers. Everything was free. Everybody got gloriously soused, including the ladies, their gentlemen callers, the cello, guitar, violin and flute."[30]

Violence, moreover, erupted sporadically in the streets. Teenage gangs warred relentlessly. Russian Jews were so frequently surrounded by "grown ruffians who . . . called them vile and obscene names, pulled their hair and beards, and . . . assaulted them with refuse, stones and clubs," that they formed a Hebrew Protective Society to ward off the attacks. And miniature "race wars" recurred with disturbing frequency. Race tensions ran so high that the mayor banned the showing of a movie of the prize fight in which a Negro, Jack Johnson, defeated a white, Jess Willard, lest the exhibition ignite a race riot.[31]

Under these conditions life in the West End became a trial for all and was virtually unendurable for the newcomers. Wholly ignorant of the ways of the city, recent arrivals had to be shown how to buy, how to cook, how to make their homes livable, and how to find employment. Many, "utterly friendless and discouraged," lamented one charity worker, succumbed to "the damnable absence of want or desire" and "grew indifferent to their own elevation."[32]

But soon outsiders saw another process developing in the West End. Boris D. Bogen, director of the United Jewish Charities and a supporter of the settlement house movement, noted that many of the greenest newcomers were "ardent orthodox, enthusiastic nationalists or earnest socialists" who "associate with their class" and "form societies, groups and congregations." Unfortunately, he felt, this outburst of community was typically short-lived and inspired only by poverty-bred desperation. After acquiring citizenship and gaining an economic foothold, Bogen complained, the immigrant

> begins to be more sceptical. . . . The struggle for existence leaves little room for any other serious interest but his own narrow, personal sphere. He becomes indifferent to social conditions . . . [and] the Settlement loses him. He joins lodges, clubs, and so on. The professional politician takes hold of him.[33]

Consequently, the Settlement developed programs "to shock inertia and foster a healthy and stimulating discontent."[34]

Yet Bogen and his allies sought more than to create reformers in the ghetto. The arrival of the overwhelmingly Orthodox Russian Jews had traumatized the city's Reform Jews. Their first response was to ridicule and shun the bearded strangers. Indeed, the precipitate flight of the Reform element from the West End succinctly dramatized this immediate reaction. But the coming of the Orthodox Russians also generated a mild wave of indiscriminate anti-Semitism which threatened to engulf all Jews. To offset this, the prosperous Reform element set out to Americanize their recently arrived coreligionists.

As a result, some Reform Jews came down from the hilltops into the ghetto. They provided jobs, loans, and social welfare institutions in their efforts to uplift the Orthodox group. To overcome the suspicion with which they were greeted, the Reform uplifters tried to create an identity between themselves and the newcomers, to manufacture a community out of Basin immigrants and hilltop Americans. To accomplish this the leaders of the Reform uplift movement had to deflect the drift of Reform Judaism toward American Protestant standards of religiosity and revive a sense of Jewishness, to create a sense of community which would cut across the division between Reform and Orthodox, the uprooted and the acculturated, the slums and the suburban fringe.[35] In the final analysis they were in league with the professional politicians. Both sought to ease the eastern Jews away from their nostalgia for Old World ways and their preoccupation with Old World issues.

These forces, combined with the Orthodox ability to take economic advantage of their imposed concentration in and around Queensgate II, got the job done. By 1914 Behr Manichewitz, the matzos king, had moved to Walnut Hills. Although 4 years later 55 percent of the city's 19,000 Jews still lived in the West End, the second exodus was gaining momentum. By 1930, out of a total Jewish population of 20,000, only 5 percent remained in the West End.[36]

Throughout these years, white educational, religious, and social welfare institutions and public buildings shifted their orientation as the population changed. Hughes High School, established at Fifth and Mound in midcentury, moved to a hilltop site at Clifton and McMillan in 1910, and for 40 years the West End had no public secondary facility (Taft High School, on Lincoln Park Drive, opened in 1951).[37] The Female Seminary at Seventh and Mound closed in the 1890s, and its building subsequently became the Pulte Medical College and, until 1926, the Ohio College of Dental Surgery. The Laura Women's Medical College and Hospital, a Presbyterian venture established in the 1890s, closed in 1904. And in 1915 the Cincinnati General Hospital moved its patients from Twelfth and Central Parkway to a new hilltop complex.

During their stay in the West End, moreover, the East European Jews significantly altered the institutional pattern of the region. In 1906, the Reform Germans at the southeast corner of Eighth and Mound moved to Reading Road and an Orthodox group, Kneseth Israel, took over the structure. That congregation—Kneseth Israel—later turned the building over to the Beth Tefila congregation, but by the 1930s the building was used only for memorial services for the dead. Two other synagogues also appeared during this period. The United Roumanian Hebrew Congregation built a small synagogue at 418 Clark Street and, in the 1930s, left it to another congregation. And in 1915, yet another Orthodox congregation opened a new synagogue on Wesley Avenue adjacent to the Wesley Female Seminary between Court and Clark.

From the late 1870s to the turn of the century, moreover, the Reform uplifters established a variety of social agencies in West End buildings. These united in 1896 to form the United Jewish Charities of Cincinnati, and the organization occupied a large structure at 231 West Sixth. Three years later the Jewish Social Settlement established itself just north of Queensgate II at 415 Clinton near Central Avenue.

The chronic instability of the land use pattern reflected the dramatic and varied changes in the composition of the population. Yet in retrospect the most important single trend was the influx of blacks. With each decade their numbers mounted. Between 1880 and 1910 the city's Negro population rose from 8,179 to 19,639. In 1920 it stood at 30,079, and in the middle of the 1920s it was estimated that the total came to 38,000, some 30,000 of whom lived in the West End. About 3,000 lived in the Fifteenth Ward, a district which included the traditionally residential part of Queensgate II above Richmond and Ninth. But the Eighteenth, covering the southern half of Queensgate II down to the river, contained over 6,500 blacks in 1920.[38]

The virtually simultaneous invasion of blacks and eastern Jews and the flight of more prosperous residents created a crisis among the Protestant churches. Many of the wealthiest simply pulled out. The Unitarians abandoned Eighth and Plum and headed for Reading Road in 1887, leaving their church to be occupied by shops and law offices. The Fifth Presbyterian at John and Clark lasted until 1913. That year the property was put up for sale and eventually passed into the hands of the St. Luke (Negro) Baptist congregation.[39]

But the crisis inspired a sense of community in some white Protestants, and they moved out more reluctantly. A bitter fight developed over the fate of St. Paul's Methodist Episcopal Church at Seventh and Smith. By 1917 the Board of Home Missionaries in Philadelphia had concluded that the changes in the neighborhood warranted the liquidation of the old congregation. The Board suggested that the church should either be abandoned altogether or turned into an institutional church (social center) under the direction of the Women's Home Missionary Society. Either decision would probably have encouraged black settlement in the area.

Most of those directly involved disliked the alternatives. The manager of the L. B. Harrison Hotel, the *Enquirer,* and local real estate men urged that St. Paul's continue as a "white man's" church in the hope that such a bastion would halt the black invasion below Ninth Street. Colored clergy in the area, however, asked that the church be given over to the Negroes. Finally, a special meeting of local Methodist Episcopal clerics announced that St. Paul's could not continue "as an independent white man's church" unless it received $2200 from outside sources. This was arranged and, when the legal question regarding the power of the national or local authorities to determine the fate of the church was decided in favor of the local officials, the question was resolved. St. Paul's remained white and the West End blacks were denied a badly needed neighborhood center.[40]

Still, St. Paul's days were numbered. In 1925 the Reverend Richard Scully conducted a residential survey of his congregation. He counted 1500 members, only 300 of whom lived in the West End. Over the vigorous objections of the 300, he recommended that the holding action be terminated, and shortly thereafter a regional church body decided to permit the building to be turned over to the Methodist Union for Negro settlement work. St. Paul's became Calvary M. E. Church and, with a black congregation of 1700, shortly emerged as the "largest Negro Church in Cincinnati."[41]

Given these circumstances it is not surprising that the Ku Klux Klan made a strong appeal to the remaining middle- and lower-middle-class white Protestants of the West End in the 1920s. Although precise membership statistics on the secret order are not available, it is estimated that membership reached 15,000 in Cincinnati during the early 1920s. Other evidence, moreover, suggests that the West End and Queensgate II especially made a disproportionate contribution to the membership rolls of the hooded society. The most outspoken clerical supporters of the Klan, most notably the Reverend Orval W. Baylor of the Richmond Street Christian Church, worked out of sites in or adjacent to Queensgate II. Fear and insecurity bred by rapid population shifts and the clash of blacks with lower- and middle-class whites, in short, provided the foundation on which a sense of community could be constructed in a specific geographic locale. Klan leaders recognized this fact and exploited it fully.[42]

Since Catholicism was one of the targets of the Klan, Catholic clergy were among the most vocal critics of the organization.[43] But the West End Catholic churches and related institutions were also in trouble in the years between 1880 and 1930. Indeed, by the mid-1890s the number of families attending Holy Trinity (German, at 619 West Fifth) had dropped to 275 from a high of 600 in the 1880s. Another German church, St. Anthony's on Budd Street just west of Queensgate II, lost 100 families in the same period, and yet another German parish, St. Joseph's at Linn and Lincoln Park, only managed to hold steady at 850 families. Interestingly enough, the congregation of St. Edward's, established in 1864 on Clark near Mound to serve the Irish, was the only Catholic church in the region to show an increase between 1880 and 1900. But after the turn of the century it, too, began to decline.[44] By the 1920s a wholesale evacuation of white Catholics was on.

Meanwhile, between 1900 and 1920, the Church responded to the northward drift of blacks into Queensgate II by perpetuating the separate but equal doctrine it had adopted in the mid-nineteenth century. The first Catholic facility for Negroes has been established in 1865. Originally located on Longworth between Race and Elm, it moved, in 1873, to a new site in close proximity to the burgeoning Bucktown ghetto beyond Broadway along east Sixth and Seventh near the Deer Creek bottoms. In 1908, when the expansion of the West End black population began to encroach on Queensgate II, St. Ann's returned to the West End on John near Richmond and Court, despite the fact that it was now archdiocesan policy to promote Americanization by gradually eliminating the ethnic parish structure.[45]

For a fleeting moment in the mid-1920s it appeared that diocesan authorities were going to abandon the separate but equal policy in favor of integration. In October of 1925, a new archbishop assured the blacks that the Church was their friend and issued a specific invitation to Negroes living in the vicinity of Holy Trinity to attend services there.[46] In fact, however, both Holy Trinity and St. Edward's had by then become rotten boroughs, and Holy Trinity itself, before the fall of 1925 was out, had become a de facto Negro parish. The transformation was officially recognized when an order of nuns devoted to the education of Indians and Negroes was brought in to staff the new school at Holy Trinity.[47]

In subsequent years other Catholic churches and institutions followed suit. The Notre Dame nunnery and academy at Court and Mound, established in 1868, was closed and replaced by Madonna High School, a coeducational facility for Negro youth. In 1929, St. Anthony's Church went black. Finally, in 1938, St. Edward's was combined with St. Ann's to create a larger social center for blacks and the archbishop instructed white communicants to henceforth attend services at St. Peter's.[48]

In the years between 1910 and 1930, then, a black neighborhood was taking shape in and around Queensgate II. Once again, census statistics provide a crude picture. By 1920 Negroes constituted about 40 percent of the total population of the region.[49] Two decades later census tracts four and eight, forming a boot along the Ninth Street and Central Avenue axis, were 90.9 and 60.3 percent black, respectively.[50] It is important, however, to recognize that, like the white neighborhoods that preceded it, black Queensgate II was diverse, divided, and mobile.

Land use patterns and the occupational structure of the region illustrate the variety. Central Avenue continued to function as the major commercial thoroughfare, though it, along with parts of Fifth and Sixth Streets, was heavily encroached upon by the expanding gambling, crime, night club, and tenderloin district which grew out from its late-nineteenth-century location along George and Longworth. A variety of black social welfare institutions developed in the old institutional core below Ninth in the southeast corner of the region. Auto repair shops moved into the old commercial and industrial quarter around this area. And the principal residential section, as in the days before the white exodus, lay above the Richmond-Ninth Street line.

The bulk of the inhabitants were poor, unskilled laborers, many of them recently arrived from the South. The newcomers were both heckled and shunned by local blacks of longer residence, and members of the two groups seldom lived side by side.[51] There were, in fact, many substantial citizens in the neighborhood, and they, apparently, tended with the passage of time to retreat northwest before the waves of migrants. One check in the 1920s revealed 202 Negro property holders in Queensgate II. But only 36 of them had lived in the neighborhood in 1900, and 76 in 1910.

Though the evidence is inconclusive, it appears that few of the city's black businessmen or professionals in the 1920s owned property in Queensgate II. There were, for example, at least 17 Negro physicians in Cincinnati in the mid-1920s, but not one appeared in the sample of Queensgate II property owners. The most common occupations in this group were barbers, caterers (or cooks), and porters. This alignment doubtless reflects the region's close proximity to what might be called the black central business district and to the city's railway terminals. It also helps explain the undertow of enmity between the West End and the more prosperous Negro district in Walnut Hills.[52] This antipathy is now part of the city's folklore, and it may in fact be related to the current distaste among some West Enders for moving to the Walnut Hills-Avondale ghetto.

The region's dense organizational structure provides another indicator of the neighborhood's diversity. The majority of West End Negro associations in the 1920s fell into the late-nineteenth-century mold, a form virtually identical to that characteristic of the city's whites. In those years, as the metropolis expanded and magnified the mobility and spatial dimension of urban life, blacks and whites alike herded themselves into a vast array of segregated and more or less ephemeral churches, secret orders, social clubs, and social-political groups. These kinds of associations continued to proliferate after 1910, although the number of black political clubs encouraged and partially sustained by whites declined in part because of the white fear of the power potential embedded in the city's swelling ghettos.

More significantly, a new kind of Negro organization, the civic and reform group, began to multiply in the West End. These, too, had their white counterparts, namely, in the network of neighborhood improvement, philanthropic, political-reform, and civic associations originated and sustained by residents of the suburban fringe after 1880. But the two sets of institutions were not identical. The whites tended to focus around and feed upon a strong sense of neighborhood pride and city boosterism. The black groups, to be sure, sought to nourish a sense of the importance of neighborhood, but in their formula urban boosterism was replaced by a vague and variously defined concept of race uplift, race pride, and black power.

The first weak beginnings of this movement began to appear in the 1890s as a response to a growing feeling among middle-class blacks that the color line was being drawn ever tighter. Segregation and caste thinking, of course, had always been a common practice and attitude among whites of all classes in Cincinnati. But after 1880 the process of urbanization gave white racism a visibly unavoidable spatial dimension. That is, the expansion of the city promoted an unprecedentedly precise sorting out of the races which resulted in a distinct and rigid physical pattern of segregation penetrating every facet of daily urban life. This had the effect of turning the black bourgeoisie back into the inner-city ghetto, within which they sought to create, temporarily at least, a fully articulated black culture.

The race pride movement, nonetheless, did not gain full momentum until after 1910. The explanation for this lies, to a significant extent, in the fact that after 1880 black society was virtually decapitated. From 1880 to 1900 the tightening color line restricted black entrepreneurs and professionals to contacts with their own race, a concentrated though overwhelmingly poor and relatively small group. The Negro middle class simply lacked the members, time, and financial resources to launch an effective organizational program to inculcate race pride and encourage social justice. After 1910, however, the burgeoning black neighborhoods in the West End changed all

that. The ghetto spawned a distinctive black bourgeoisie composed of under-
takers, theater operators, porters, doctors, teachers, preachers, and social
workers who possessed the time, energy, and organizational talents to give
black power a voice in Cincinnati.

But they were not alone, for the Klan constituted only one white response
to the blackening of the slums. White civic reformers from the suburban
fringe—a minority, to be sure, but a significant one—had since the 1880s
been vitally interested in the fate of the center of the city. After 1910, many
of the new black spokesmen, including Wendell P. Dabney, editor of a
newspaper called *The Union* and a student of Cincinnati's black history,
sought out and found allies among that group. It was this peculiar combination of circumstances that made the roaring twenties the heyday of the black
West End and a decade of hope and paradox.

A clear majority of the new Negro organizations located in that part of
Queensgate II below Ninth which had been the region's commercial, industrial, and church district in the 1870s. There was at least one agency for each
pressing problem. A colored YMCA and YWCA sought to house and educate Negro Youths, as did the Home for Colored Girls and the Evangeline
Home for "unfortunate colored girls," a joint undertaking of an integrated
group of Cincinnati philanthropists and the Salvation Army, and the Friend-
ship Home for Colored Girls, a Women's Home Missionary Society enterprise. The Negro Civic Welfare Association, affiliated with the citywide
Council for Social Agencies, and the predecessor of the local Urban League,
worked to encourage social work among Negroes, surveyed social conditions in the West End, supported model housing projects for blacks, helped
improve recreational facilities, established day nurseries, promoted public
health, and fought job discrimination in the city's industries.[53] The Colored
Industrial School, established in 1914 through the efforts of a white woman
philanthropist, offered regular courses through the tenth grade and a special
three-year trade course especially designed for those who had to drop out of
school to earn a living.[54]

If Dabney's history of Cincinnati's Negroes provides any attitudinal
guide, moreover, most of the black participants in this movement were
stimulated in large part by suspicion and dislike of a white community which
simultaneously ghettoized and patronized the city's blacks. These Negro-
white alliances, if such they were, provided the blacks a chance to use white
resources as a means of uplifting ghetto residents and disproving the caste
assumptions that dominated white thinking.

Nonetheless, a large though incalculable number of Queensgate II Negroes decided that the best strategy for the attainment of race pride and black power was to ignore the whites and concentrate on the development of black culture and entirely black institutions. Clearly this policy had a broad appeal. Even Dabney, who was always willing to work with whites, responded to the magnetic pull of black independence by writing a history of Cincinnati Negroes which, though not uncritical of his brothers, focused on the record of black achievement in the face of overwhelming obstacles.[55]

Others in the 1920s struck out less ambiguously in the same direction. One institutional manifestation of this was the Cosmopolitan School of Music, established in 1921 at 823 West Ninth Street. Claiming to be the only fully accredited Negro owned and operated school of music in the United States, it concentrated its instruction on music, expression, and language, and was especially dedicated to the perpetuation and advancement of Negro art and music.[56] The other, and clearly the strongest, black cultural organization was the Cincinnati branch of the Universal Negro Improvement Association and African Communities League. Its president in the 1920s was William Ware, a resident of 927 Barr Street and native of Lexington, Kentucky, who came to Cincinnati in 1903. In 1917 he founded and served as the first president of the Welfare Association for the Colored People of Cincinnati. In August of 1920, however, after attending the first International Convention of the Negro Peoples of the World in New York City, he returned to Cincinnati and reorganized the Welfare Association into the UNIA. By the mid-1920s the local group had enrolled 8000 members.[57]

Though no other organization in the West End approached the aggressiveness of the UNIA in fostering black pride, the same though somewhat muted overtone appeared in yet another relatively new type of black organization, the neighborhood improvement association. Several of these groups appeared in the West End in the 1920s, including a Baymiller Improvement Association, a Central Avenue Improvement Association, and the African Improvement League located on George Street. Their principle concern was the "unfavorable public image" generated by an area rapidly becoming infamous for its overcrowded and dilapidated housing and outrageous crime and vice record. Their efforts, abetted by the work of Dabney as a liaison to influential whites, the *Cincinnati Post,* and several elected officials, spearheaded a crusade against the inadequate residential, health, and recreational conditions in the West End.

Much of energy behind this drive was directed toward goading the city into providing expanded recreational facilities. The attack continued throughout the 1920s, but City Hall, despite the adoption in 1924 of the city manager system coupled with proportionate representation, was hardly responsive. Although by 1927 the city council had allocated a total of $570,000 for recreation in the area, only $150,000 had been spent.[58]

Despite this frustration, the West End and Queensgate II in the 1920s hummed with activity and buzzed with a fretful optimism. Though racked with all the problems of the worst sections of the twentieth-century metropolis, its people knew they were alive. Protestant, Catholic, and nonsectarian white philanthropists had begun to focus their attention in it, the urban Klan found it a fruitful frontier of organizational activity, and the black bourgeoisie, though divided, was beginning to hack out diverse routes to community improvement, cultural development, influence, and prestige in the affairs of the city.

The vice, crime, and gambling of the tenderloin made the region both a terror and an avenue of escape from the grim realities of a segregated urban system. And violence between blacks and whites on the one hand and Negro newcomers and veterans of the ghetto on the other bubbled ominously near the surface. Though poverty was the great common denominator of the section, an occupationally and residentially defined class structure had begun to develop. And, despite the low economic level of life, the concentration of blacks and the region's mix of employment opportunities made it possible for some to eke out a tolerable existence and for a few others to reach a level of comfort somewhere above the minimum. Queensgate II and most of the West End, to be sure, was a slum, but it was neither stagnant nor yet doomed to indefinite decline. A balanced assessment of the forces of flux in the 1920s would have to conclude that options remained, that the vital undertow of optimism was not unwarranted. Though not a community, Queensgate II was still part of a viable neighborhood.

The Emergence of an Area

By the mid-1960s that same neighborhood constituted an indefinable geographic space identifiable only as an area slated for urban renewal and dubbed Queensgate II. Its population was appallingly poor, disproportionately old, and discouraged by the prevalence of petty criminality and drunken dereliction. The small middle-class residue was terrified not so much by the disarray of their physical and human surroundings as by their lack of control and the prospect of forced removal into a new neighborhood. Firsthand and lifelong acquaintances with the consequences of uprootedness, they faced with dread the prospects of becoming strangers in their own city. Consequently they organized and insisted, through the West End Community Council and Queensgate II Club, the only channels open to their grievance, that their neighborhood be at least preserved if not revived. Desperately afraid, they could not afford the luxury of a riot. Their fate, and that of the rest of the area's inhabitants, in the final analysis, rested with the outsiders who met regularly in City Hall, St. Peter's, and the Old Vienna, a restaurant across the street from City Hall.

In retrospect, though all the evidence is not yet in, it is not difficult to discern some of the major reasons for the transformation of Queensgate II from a lively slum into a dying if not dead neighborhood. Though poverty, blight, crime, demographic change, and overcrowded and unsanitary living conditions persisted in Queensgate II proper, these ominous forces, so terrifying to both lay and professional students of the city, were neither new nor principally responsible for stifling the vitality of the slum of the 1920s. It was, instead, choked off by a passion for open spaces, industrial development, and, paradoxically enough, urban reform and improvement. Combined, the changes wrought by these programs drastically altered the historically flexible physical structure of the region, reduced the level of diversity, and cut Queensgate II's life-sustaining connections to its larger neighborhood.

The process began in the 1930s. By 1933 Lincoln Park, located at the end of Laurel Avenue just north of Queensgate II, had been replaced by a new Union Railway passenger terminal. To facilitate transportation between it and the central business district, the city brought up property and constructed a beautifully green and very broad boulevard, Lincoln Park Drive. With the completion of this project a wall had been started to the northwest of Queensgate II, and depopulation began.

Next came public housing. On December 16, 1935, the demolition of 350 dilapidated dwellings north of old Lincoln Park began. The Laurel Homes project was completed by 1938, and in 1941–1942 the second project, Lincoln Court, located directly north of Queensgate II and bounded by John, Clark, Cutter, and Lincoln Park Drive, went up. The wall to the north and northwest was all but complete, and the construction of Taft High School in 1951 on Lincoln Park between Central and John finished the job.

After World War II, attention focused on the notorious slums immediately to the south and southwest of Queensgate II. But now the objective was not public housing or recreation, but industrial development and transportation. The automobile and truck operating on the limited-access expressway and the inner-city industrial park were now regarded as the instruments that could extricate the city from the postwar urban crisis. The intramural industrial park southwest of Queensgate II had strong appeal because of its contiguity to downtown and because, if successful, it would bolster sagging city revenues. The expressway, moreover, not only provided modern rapid transit for the Mill Creek industrial belt but also facilitated the travel of commuters into and out of the post-World War II metropolis. An earnings tax mitigated the financial shock administered by the cost of road construction and the consequent expansion of slums and increase of welfare expenditures within the central city's corporate boundaries.[59]

In the 1960s both these projects were completed and, with the central business district on the east, finished the barricade around Queensgate II. The effect was devastating. The altered pattern of land use brought on by the burst of improvements drastically depopulated the entire West End. In 1960 the total stood at 41,949 and by 1966 it had dropped to 29,137 as Walnut Hills-Avondale emerged as the city's most recent black ghetto. Queensgate II was hurt even more seriously. In 1960 it contained but 4,687 souls; by 1966 that had dropped to 2,255; two years later it sank to 1,280.

Conclusion

From the perspective of the urban historian, Queensgate II is not now and never has been a community. But it has a long and rich past as part of a larger neighborhood, the West End, the experience of which has been irrevocably tied to the forces and individual and group decisions that have determined the course and patterns of urbanization for over a century. It was, successively, a semirural suburb, a marginal region on the urban fringe, a heterogeneous but respectable residential district, and, into the 1930s, a central-city slum. In each of these stages, moreover, it was characterized by flux and diversity in land use patterns, occupational structure, population composition, and manifestations of the unfulfilled desire for community. It began to die in the 1930s because it was cut off from the larger neighborhood of which it was a part. Gradually businesses, schools, welfare institutions, and people left the area. Its functions and population became increasingly homogeneous. By most criteria it had always been a more interesting and promising place in which to live than it was in 1960. Historically a staging ground for mobility, it is now a dying if not a dead neighborhood.

But it need not remain a captive of history. No reasonable person, of course, would suggest that Queensgate II be reconverted to the slum of the 1920s, although that would be a significant improvement over the present. Fewer still, doubtless, would try to return it to its precise status and condition as of 1830, 1850, or 1880. But surely we can benefit by the obvious lesson of history. The neighborhood throve on diversity, for it was diversity which gave the area flexibility and helped nourish change, and hence the hope of improvement. There is no reason, other than the lack of will and foresight, that it cannot be reconstituted into a diverse neighborhood (regardless of its racial composition) capable once more of generating the enthusiasm and power to play a significant role in determining its own ultimate fate and the future of the city. History is instructive here, too, for even this brief review makes it clear that in a live neighborhood virtually anything is possible. But a vital neighborhood, like a great metropolis, has never been built by people of little courage.

NOTES

1. The Betts house, a plain, two-story, brick structure, is among the oldest remaining houses in the city.

2. See Louis Boeh, "Queensgate II: The Historical Development of a Segment of a Community, 1800-1845," pp. 5-6, Urban Studies Collection, Archival Collections of the University of Cincinnati (hereafter cited as USC, UC).

3. Between 1830 and 1850 the city's central commercial district moved from its original focus in the Ohio River bottoms along Front and along Main Street up to Second, Third, and Fourth between Broadway and Elm. See Central Business District Map, 1830-1875, USC, UC; and Charles T. Greve, *Centennial History of Cincinnati,* Vol. 1 (Chicago, 1904), pp. 349-350, 418, 556, 675.

4. For the fate of the cemeteries, see Boeh, "Queensgate II: Historical Development," pp. 5-7.

5. The Seventh contained 2356 males and 2384 females in 1840. Almost one-half of the total were 30 years old or younger, and the largest single age group was the 20-30 category. Charles Cist, *Cincinnati in 1841* (Cincinnati, 1841), pp. 32-33.

6. The following table shows the construction pattern in the West End above Sixth from 1840 to 1844. We do not, unfortunately, have figures for the period to 1840.

Year	Brick Construction	Frame Construction	Subtotal
1840	44	54	98
1842	145	604	749
1843	138	51	189
1844	120	106	126
Totals	477	815	1292

See Cist, *Cincinnati in 1841*, p. 42; also Charles Cist, *The Cincinnati Miscellany, or Antiqui-ties of the West; and Pioneer History and General and Local Statistics*, Vol. 1 (Cincinnati, 1845), p. 114.

7. See the 1850 Land Use Maps, USC, UC.

8. Sherry O. Hessler, "'The Great Disturbing Cause' and the Decline of the Queen City," *Historical and Philosophical Society of Ohio Bulletin* 20, 3 (July 1962): 169-185. See Street Railway Map, 1878, USC, UC; Greve, *Centennial History of Cincinnati*, pp. 743-744; Moses King, *King's Pocket Book of Cincinnati* (Cambridge, 1879), pp. 41-42.

9. See Central Business District Map, 1830-1880, USC, UC.

10. Although before this period Negroes were scattered along the waterfront, the most significant concentrations in the 1820s were in two wards flanking the central waterfront. See Richard C. Wade, "The Negro in Cincinnati, 1880-1830," *Journal of Negro History* 39, 1 (January 1954): 44.

11. *Enquirer*, September 30, 1852; *Commercial*, May 13, 1865; *Commercial-Gazette*, May 24, 1896; *Enquirer*, August 22, 1869; Federal Writers' Project, *A Guide to the Queen City and Its Neighbors* (Cincinnati, 1943), pp. 195-196.

12. D. J. Kenney, *Illustrated Cincinnati: A Pictorial Handbook of the Queen City* (Cincin-nati, 1875), p. 94; D. J. Kenney, *Illustrated Guide to Cincinnati* (Cincinnati, 1893), pp. 160-161, 165-166; Zane L. Miller, "Boss Cox and the Municipal Reformers: Cincinnati Pro-gressivism, 1880-1914," Ph.D. thesis, University of Chicago, 1966, ch. 3.

13. For these and other developments within the Jewish community see Barnett R. Brick-ner, "The Jewish Community in Cincinnati," Ph.D. thesis, University of Cincinnati, 1933.

14. Quoted in Louis Boeh, "Queensgate II: The History of a Segment of a Community, 1845-1895," p. 5, USC, UC.

15. The original location of St. Paul's was at Fourth and Plum. The Second Presbyterian originally stood on Fourth between Vine and Race. Cincinnati *Daily Gazette*, November 19, 1870; King, *Pocket Book of Cincinnati*, p. 73.

16. Kenney, *Illustrated Cincinnati*, pp. 107-109; Kenney, *Illustrated Guide to Cincinnati*, pp. 127-129; King, *Pocket Book of Cincinnati*, p. 33; *Times-Star*, August 11, 1933, p. 22; *Commercial Tribune*, April 5, 1924, p. 3.

17. Miller, "Boss Cox and the Municipal Reformers," p. 5.

18. See Land Use Maps, 1875, USC, UC: Historic Landmark Map, USC, UC; Miller, "Boss Cox and the Municipal Reformers, p. 19.

19. There were shopping districts on Cutter, Linn, and Freeman, but their density decreased as one moved west.

20. Richard Nelson, *Suburban Homes* (Cincinnati, 1874), pp. 8-9.

21. Zane L. Miller, *Boss Cox's Cincinnati: Urban Politics in the Progressive Era* (Chi-cago: University of Chicago Press, 1980; Westport, CT: Greenwood, 1981), Introduction.

22. Miller, *Boss Cox's Cincinnati*, chs. 1-3.

23. Quoted in Miller, "Boss Cox and the Municipal Reformers," pp. 19, 32.

24. *Catholic-Telegraph*, August 14, 1884, p. 1; *Times-Star*, ct. 19, 1892, p. 8.

25. Quoted in *American Israelite*, October 26, 1899, p. 6.

26. Thomas Stanley Matthews, *Name and Address* (New York: Simon & Schuster, 1960), pp. 3, 8, 26, 27.

27. Miller, "Boss Cox and the Municipal Reformers," p. 27.

28. Miller, "Boss Cox and the Municipal Reformers," pp. 41-42, 70, 136.

29. Miller, "Boss Cox and the Municipal Reformers," p. 20.

30. Oscar Ameringer, *If You Don't Weaken* (New York: Holt, Rinehart & Winston, 1940), p. 49.

31. Miller, "Boss Cox and the Municipal Reformers," p. 54.

32. Miller, "Boss Cox and the Municipal Reformers," p. 35.

33. Quoted in Miller, "Boss Cox and the Municipal Reformers," p. 50.

34. Miller, "Boss Cox and the Municipal Reformers," p. 50.

35. Miller, "Boss Cox and Municipal Reformers," pp. 524-567.

36. Barnett R. Brickner, "The Jewish Community in Cincinnati," pp. 25, 27, 442.

37. Federal Writers' Project, *Guide to the Queen City*, p. 373; *Times-Star,* December 2, 1955.

38. Wendell P. Dabney, *Cincinnati's Colored Citizens* (New York: Johnson Reprint Corp., 1970; originally published in 1926, by the author), p. 382; U.S. Bureau of the Census, *Fourteenth Census, Vol. III, Population* (Washington, DC: Government Printing Office, 1922), pp. 799, 800.

39. *Commercial Tribune,* February 7, 1913; Seven Hills Neighborhood House, *A Project for Protection, Clark Street: A Community Project Report* (n.p., n.d.), p. 13.

40. *Commercial Tribune,* July 10, 1917; *Enquirer,* August 2, 14, 16, 27, 1919.

41. *Commercial Tribune,* August 29, 1925; *Enquirer,* March 31, 1957.

42. Kenneth T. Jackson, *The Ku Klux Klan in the City* (New York: Oxford University Press, 1967), pp. 164-165, 239; David Calkins, "The Cincinnati Protestant Clergy in Social, Political, and Theological Reform, 1915-1929," M.A. thesis, University of Cincinnati, 1966, pp. 47-49.

43. Jackson, *The Ku Klux Klan,* pp. 164-165.

44. Figures on the size of the Catholic congregations were compiled from the *Summary of [Church] Records, Archdiocese of Cincinnati,* located in the diocesan archives.

45. Miller, "Boss Cox and the Municipal Reformers," pp. 579-649.

46. Quoted in Dabney, *Cincinnati's Colored Citizens,* p. 376.

47. Anonymous, *Diamond Jubilee of St. Edward's Church and the Annals of St. Ann's Mission* (n.p., n.d.), p. 7. Simultaneously the archdiocese intensified the stress, dating to 1910, on social welfare work. See Louis Boeh, "Queensgate II: The Historical Development of a Segment of a Community, 1895-1930," pp. 8-9, USC, UC.

48. Anonymous, *Diamond Jubilee,* p. 7; *Enquirer,* September 14, 1938; *Post,* September 9, 1938; November 3, 1939.

49. U.S. Bureau of the Census, *Fourteenth Census,* p. 800.

50. U.S. Bureau of the Census, *Sixteenth Census Population and Housing Statistics for Cincinnati* (Washington, DC: Government Printing Office, 1942), p. 4. Also see "1940-1960 Census Tract and Block Map, Queensgate II," USC, UC.

51. Interview, Henry Richter, March and November 1967.

52. This rivalry is expressed in more or less subtle remarks scattered through Dabney's book.

53. Dabney, *Cincinnati's Colored Citizens,* pp. 211-233.

54. Dabney, *Cincinnati's Colored Citizens,* pp. 355

55. Ironically, the white booster-reformers of the 1880-1920 period used the same technique—that is, the combination of urban history and biographical record of accomplishment—to foster a sense of devotion to place which would serve as stimulant to civic improvement and municipal reform.

56. Dabney, *Cincinnati's Colored Citizens*, pp. 212-213.

57. Dabney, *Cincinnati's Colored Citizens*, pp. 213-214. The objects of UNIA were "to establish a confraternity among members of the race; to promote the spirit of pride and love; to reclaim the fallen; to administer to and assist the needy; to assist in civilizing the backward tribes of Africa; to assist in the development of independent Negro nations and communities; to establish commissionaries or agencies in the principal countries and cities of the world for the representation and protection of all Negroes . . . ; to promote a conscientious spiritual worship among the native tribes of Africa; to establish universities, colleges, academies and schools for the racial education and culture of the people; to conduct a world-wide commercial and industrial intercourse for the good of the people; to work for better conditions in all Negro communities" (quoted in Dabney, *Cincinnati's Colored Citizens*, p. 214).

58. *Times-Star*, August 12, 13, 1927; Federal Writers' Program, *Guide to the Queen City*, p. 231. Also see Cincinnati, *Municipal Activities, 1933* (Cincinnati, 1933), pp. 33-34.

59. *Enquirer*, January 30, 1946; *Post*, June 5, 1961.

PART II

Community Perspectives

The essays in this section assess community participation in the Queensgate II project from three distinctive viewpoints. Thomas H. Jenkins, director of the university planning and research team during the project, approaches the subject as a participant, a planner, and as a student and teacher of urban planning and sociology. For him, community participation was a fact of life in the 1960s and a "given" in the Queensgate II project. Moreover, he does not take the metropolitan community, the city, or the neighborhood of Queensgate II as the community of concern. Instead, he defines community as an entity composed of the various institutions with a direct interest in the fate of Queensgate II, the residents of the area, and those who chose to help define and defend that residential interest. For him, therefore, community participation means the coordination through the West End Task Force of both the institutional and residential actors who made up the Queensgate II community. For him, too, such coordination ought to use a human approach to planning which places human interests on an equal plane with institutional interests. In retrospect, Jenkins finds that representatives of the residents had the most influence on Task Force deliberations and the plan itself, and that as those representatives moved out of the action during the early implementation stage the planning process worked differently and to the detriment of the vision embodied in the plan as presented to the Task Force.

Richard Lewis and Jerome Jenkins, both of whom were members of the Task Force, speak next and from a different perspective on community. Ostensibly representatives of those who lived in Queensgate II, they saw themselves as leaders in "the movement" that aimed to build a stronger and more effective sense of community among blacks generally in Cincinnati as a means of improving the lives of the individuals who made up that community. For them, therefore, the Queensgate II project was both a substantive and a symbolic issue in a long-range drive for human rights in a community they define in racial rather than in geographic or institutional terms.

The last community perspective comes from Edgar J. "Buddy" Mack, who chaired the Task Force before and during the Queensgate II planning process. Mack reflects on his experience in that post, with special reference to the role of the Task Force in Queensgate II planning and to his strategy as head of the Task Force. He presents himself as a good citizen fulfilling his obligation to improve the quality of life of the individuals who lived in the Cincinnati metropolitan area. And for him, quality of life meant not only access to decent housing, education, and jobs for impoverished people from the inner city, but also the survival and vitality of institutions of amenity, such as Music Hall, which serve individuals from across the face of the metropolitan area.

These essays, then, can be seen as doing several things. They provide assessments from differing vantage points of how the planning process worked. They measure, by varying standards, the utility of the process in achieving varying goals. And they tell us without saying so that the definition of community among these participants differed in particulars, but that all of them took the welfare of individuals and/or individual institutions, rather than that of the metropolis, the city, or neighborhood community, as the basic unit of concerns. They do not speak, that is, to the "public interest" as something attached to a geographic community, such as the metropolis, city, or neighborhood. These essays, therefore, form part of a discourse based on a set of assumptions about the fundamental units of urban society which appeared in the mid-1960s and which has persisted into the present.

CHAPTER 3

The West End Task Force:

COMMUNITY PARTICIPATION AND POLICY PLANNING

Thomas H. Jenkins

Official recognition of both citizen participation and coordinative planning in the Queensgate II project entered by way of the West End Task Force, a group appointed in 1966 with the approval of the city council by the city manager.[1] The authorizing language described the Task Force as a "community task force" empowered to review and supervise all planning and development activities in the West End of Cincinnati, including the Queensgate II project. In fact, however, its structure embodied a special definition of "community" which merged the new sensitivity among planners and city officials for the human approach, the needs of local residents, and coordinative planning among disparate institutions and groups.

The composition of the Task Force defined the "community" for planning purposes as those individuals and groups which possessed virtually any legitimate social, economic, or political interest in the future of the West End (see list of members and their affiliations, Table 3.1). Its membership included, as a consequence, an astonishing variety of people who either lived, worked, did business, or provided public or private services in the West End.

TABLE 3.1   West End Task Force: Members and Affiliations (as of 1969)

Name	Affiliations
Rev. Robert Beck	Member, West End Community Council (Pastor, First Reformed Church)
Mrs. Doris Brown	President, Queensgate II Community Club
Mrs. Marybelle S. Brown	Executive Director, West End Branch YWCA
Mrs. A. L. Davis	Member, West End Homeowners' Association
Mrs. Rose Daitsman	Member, West End Community Council
George Hayward	Member, Urban Development Committee, Greater Cincinnati Chamber of Commerce
Richard Hartke	Member, West End Business Association
Walter Hempfling	President, West End Industrial Association
R. Jerome Jenkins	Deputy Director, Seven Hills Neighborhood Houses, Inc.
Msgr. Joseph Kennedy	Pastor, St. Peter in Chains Cathedral, Roman Catholic Church
Peter Kory	Director, Department of Urban Development, City of Cincinnati
Richard W. Lewis	Staff, Seven Hills Neighborhood Houses, Inc.
Edgar J. Mack	Treasurer, Cincinnati Symphony Orchestra (Officer, Seasongood & Mayer, Investment Counselors)
Hon. William J. Mallory	Member, West End Community Council (State Representative)
James L. Rankin	Staff, Seven Hills Neighborhood Houses, Inc.

Name	Affiliations
Carl Mayerhofer	Executive Director, Cincinnati Metropolitan Housing Authority
Herbert W. Stevens	Director, City Planning Department, City of Cincinnati
Rev. Richard Sullers	Pastor, West End Presbyterian Church
Guy Buddemeyer	Staff, Cincinnati Public Schools
Mrs. Georgiana Wynn	Resident at Large

Despite the Task Force's heterogeneity, it functioned much more aggressively and effectively than traditional citizen groups; it can be classified as a supplementary decision-making body which, on occasion, acted as a citizen control mechanism. Indeed, much of its strength derived from the breadth of its membership. Task Force members had ties with a range of interest groups, including neighborhood associations, churches, small merchants, large businesses, medium-sized industries, social and public service agencies, one large utility, and, through its black members, including a state representative, both the local Democratic Party and the larger black community as well. In addition, the Task Force was "augmented" whenever it met to make recommendations to the Planning Commission by the inclusion of some city council members and city planning commissioners. These representatives of this broadly defined "community" formed a ready conduit through which influence could be exerted both on city hall and on other power points within and beyond the metropolitan area.

The structure of the Task Force proved to be an important source of its influence in other ways. It stood, for example, as a mediator between the group representing the residents of Queensgate II and the city government, providing each the opportunity to "educate" the other. Similarly, augmenting the Task Force with city councilpersons and city planning commissioners brought planners and policy makers together in a helpful manner. By the time Task Force recommendations reached the Planning Commission and Council, members of each of these bodies had participated in developing the proposals on which they had to vote as policy makers. This arrangement reduced the gap and the traditional distrust and friction between those who formulated the goals and plans and those who provided the laws, money, organization, and other tools to carry them out.

The local model for this combined participation and decision-making form and process was the Working Review Committee, a device created in 1963 to guide the redevelopment of the central business district. It included as participants in early stages of planning those involved later in the implementation of plans and policies. It held regularly scheduled meetings open to the public. It recommended goals and policies to the city administration which went on to the city council in only slightly modified form. And professional expertise was furnished it by City Hall, either in the form of loaned city staff or in the form of a paid consultant. Of these features, as Anthony Downs of Real Estate Research Corporation has said about his work in Cincinnati with the Working Review Committee, the most important for securing authentic participation was the inclusion of both planners and implementers in the process from the beginning.[2]

Additional insurance of Task Force influence came from the definition of the relationship between the university planning team and the Task Force built into the city-university contract, a contract which the Task Force itself helped formulate. The document specifically required the university team to prepare a work program and schedule designating the major decisions to be made by the Task Force during the project, to submit all proposals for such decisions to the Task Force for approval, and to present all proposals at an open meeting before the Task Force.

Another source of Task Force influence was the support given to it by the professional staff supplied by the City Planning Commission. Indeed, the Planning Commission staff and Task Force grew so close in the two years between 1966 and 1968 that key members of the Task Force displayed considerable reluctance to accept the university planning team as contractor-consultant for the Queensgate II project. They preferred the Planning Commission, a known and trusted ally, to an unknown quantity that might turn out to be too "neutral" or even "proestablishment."

Yet the structure of the Task Force and the role of Planning Commission staff do not alone account for the effectiveness of "community" participation in the Queensgate II experience. Indeed, in one respect the structure of the Task Force seemed designed to undermine the influence of one faction of the community, for only three blacks of the twenty members of the group were chosen primarily on the basis of their status as neighborhood residents: Mrs. Doris Brown, a homeowner in Queensgate II and president of the Queensgate II Community Club and the West End Community Council; Mrs. Georgiana Wynn, designated "resident at large"; and Mrs. A. L. Davis, who represented the West End Homeowners Association. Yet the apparent underrepresentation of resident citizens concealed a larger political reality. As the planning process evolved, at least nine other people actively guarded the interests of the predominantly black residential community.

The central figure for black participation on the Task Force was Democratic State Representative William Mallory. He had grown up in the West End and developed an instinctive feeling for the attitudes and aspirations of West Enders nurtured in the streets, lobbies, bars, and soul food dining rooms of the neighborhood, and in the kitchens, living rooms, and churches of his friends, neighbors, and constituents. Mallory and Mrs. Brown, along with Jerome Jenkins, Richard Lewis, and James Rankin, all blacks and all residents or former residents of the West End who worked as professional staff for the Seven Hills Neighborhood Houses, Inc., ranked among the most vociferous, militant, articulate, active, and aggressive members of the Task Force. They received steady support from yet another member of the Task Force, the Reverend Robert Beck, a white minister in a West End church who lived in Queensgate II. He frequently leveled sharp criticism at the city for its policies in the West End and, though not designated a residents' representative, served *de facto* in that capacity.

Four other residents or former residents who were not members of the Task Force also played an active and effective part in the planning. Mrs. Rose Daitsman, a neighbor of Jerome Jenkins in the West End's Park Town apartments and one of the few whites living in Queensgate II in 1968, worked through the West End Community Council's Area Planning Committee. Paul Henry, a black man of exceptional organizational skills who held a position in the community relations division of the University of Cincinnati and taught at the university, played a behind-the-scenes and unofficial role as a person to whom residents, university people, and city officials could turn for counsel and advice during crises and breakdowns in interinstitutional communications. The third, Lucille Burden, a soft-spoken black woman, Queensgate II resident, and secretary of the Queensgate II Club, functioned initially as a key lieutenant of Doris Brown's and later as liaison between the Queensgate II residents and the university planners. Finally, there was Latham Johnson, a student at Brandeis University and a former crony of Jerome Jenkins, Lewis, Rankin, Mallory, and Henry, who possessed a long record as a militant activist in the West End. He kept in touch with the project by phone and through occasional visits to Cincinnati blacks and to several of the more progressive members of the Graduate Community Planning Department at the University of Cincinnati.[3] He preached the gospel of neighborhood control and served as the intellectual-in-absentia for the residents and their allies.

Two other blacks exerted a strong influence on Queensgate II planning, though neither served on the Task Force or possessed the close ties to the West End characteristic of most of the others. Harry Martin was a social worker and a native of Cincinnati whose parents lived in Kennedy Heights, an interracial section in the northeast part of the city, who provided a kind of father figure for the black professionals involved in the planning process, and fed important ideas to those officially involved with the planning. During the early years of the project he was Model Cities coordinator in Cincinnati, and his educational and professional social work experience in the city gave him detailed knowledge of City Hall and local politics and formed the basis for his conviction that the city's blacks suffered from acute political anemia, especially within the political parties and City Hall bureaucracy. Finally, Charles Collins, a former Republican city councilman and a member of the City Planning Commission who occasionally attended augmented meetings of the Task Force, used his position as a friend of Mallory's and other Task Force members to defend his perception of the community's interest, a perception sharpened by growing up in Cincinnati and coming up the hard way.

This group in a variety of ways and in a variety of combinations defended the residents' interests in the planning process. Some of their activities and accomplishments occurred through official channels. They played a key role on the Task Force in shaping its policy decisions of 1967 for redevelopment of the West End, a document which the city and university planners recognized as the general guideline for Queensgate II planning. They participated in 1968, directly and indirectly, in a trip with Mayor Eugene Reuhlmann and others to Washington, D.C., to lobby the Department of Housing and Urban Development for the reservation of $12,700,000 of federal funds for Queensgate II. That same year they helped formulate the details of the city-university contract, including those provisions establishing the decision role of the Task Force in the process, procedures for the planners' presentations, the length of the contract, and how the planners would be paid in relation to the review of their performance.

In two cases early in the planning, those on the Task Force most directly concerned with resident interests secured action with immediate payoff for their special clientele. In 1968, Doris Brown and Jerome Jenkins pushed hard for the immediate establishment of the West End Health Clinic, a proposal which grew out of a survey of community health services conducted by the Task Force's Health Committee aided by staff from the City Planning Commission.[4] The facility, though modest, provided diagnostic service on Saturdays out of a two-bedroom apartment in a Cincinnati Metropolitan Housing Authority project on Linn Street, and it stood as a direct rebuke to a health survey of that year conducted under the auspices of City Council and the County Commissioners which contended that "poor people . . . prefer rather run-of-the mill medical treatment, which is readily available, to the very finest of medical care, which is not." And in 1969, Brown, Jenkins, and Lewis spearheaded the successful effort to secure an explicit commitment to create employment for West End residents and business for small merchants in all development proposals, a commitment Mrs. Brown never let us forget for a minute.

That same year resident representatives persuaded the Task Force to stop making decisions on the plans for over a month until settlement of a flap between the Task Force and City Hall over the handling of the Town Center aspect of the plan. The dispute erupted when the city's Director of Urban Development, Peter Kory, presented plans for the Town Center to three committees of City Council without giving prior notice to the Task Force. Jenkins, Rankin, and Collins reacted by implying that Kory's boss, City Manager Richard Krabach, had tried to bypass the Task Force in an important planning decision, a step which, in their view, violated the decision-making procedures previously agreed to by all parties in the partnership. At one point in the argument Krabach told the Task Force that "its power cannot exceed those of determining facts and making recommendations," but in the face of strenuous opposition to that position he later conceded that he saw the Task Force as a serious partner in planning and development, albeit one inhibited by technical (staff) limitations.

The Town Center hassle also centered on another issue: the substance of the proposal presented by Kory. Many on the Task Force, among them both the official and unofficial representatives of the residents, thought Kory's scheme conceived of the Town Center as a subunit of the central business district. The Task Force, however, viewed it as a regional center to serve, benefit, and focus on the greater West End. These different concepts, they contended, would affect, in a practical and vital way, rent scales, the number of expensive boutiques, the types of merchandise, and the kinds of cultural and recreational facilities in the Town Center, as well as the amount of community investment in and control of the development. The Task Force protest resulted in the shelving of Kory's proposal and in its subsequent modification through standard Task Force planning procedures.

The next contribution of the group representing the residents took place in 1969 and 1970, when Mallory initiated the creation of the West End Development Corporation (WEDCO) to oversee the construction of the first phase of the Queensgate II project.[5] The idea took shape in a special meeting in St. Luke's Baptist Church in May of 1969 as a means not only of providing West End residents with housing but also to secure them a voice in determining the number, size, location, type, and design of the units.[6] Mallory and Doris Brown were among the original incorporators, and after a series of negotiations the city named WEDCO and Mid-City Development, Inc., of Washington, D.C. (then headed by H. Ralph Taylor, a former federal official in Model Cities), as developers of the first 600-unit phase.[7] In understandings worked out in meetings among Taylor and Mallory, a few other individuals, resident groups, and the Task Force, the joint venture became a tandem arrangement, with each of the two firms sharing in the division of labor according to their special capabilities, interests, and locations.[8]

Yet WEDCO's role extended beyond the development stage, for the corporation intervened in the planning process in mid-1970, before the completion of the final plan. After reviewing the university planning team's housing guideline proposals, WEDCO brought to the Task Force a series of amendments affecting the distribution, interrelation, characteristics, and design of housing units, requirements for pedestrian access, provisions for private outdoor leisure space, the variety of housing types for families with and without children, and the socioeconomic classes to which Queensgate II housing should appeal. Thus the residents' development corporation significantly altered plans which it expected later to develop.[9]

Resident sentiment also affected Queensgate II planning through channels essentially outside the Task Force as a consequence of militant participatory activities so characteristic of the 1960s and early 1970s. In those years, as mentioned earlier, cities across the country felt the impact of an increasingly aggressive mood in the civil rights, student, feminist, and antiwar movements, and among professional athletes, school teachers, university professors, planners, and social workers. All these movements manifested themselves in Cincinnati and helped create a crisis atmosphere around the planning process.

The new militancy among blacks proved the most important for the Queensgate II experience. In June of 1967, one year before the onset of Queensgate II planning, a rebellion among blacks in Cincinnati produced three outbursts of violence, the arrest of 404 persons, and the inclusion of Cincinnati in the group of cities studied by the National Advisory Commission on Civil Disorders (the Kerner Commission).[10] Though no further violence erupted in the West End, we had to deal with other forms of black power militancy.

Our first encounter occurred during the first few months of the planning team's work and involved Father Sicking, an occasionally abrasive white Catholic priest known for his impatience with Church bureaucracy and government red tape, and a group of young West End activists known as the Black Turks. Father Sicking had resigned from the local board of the Community Action Commission in protest of what he regarded as its ineffectiveness and, as Director of the DePorres Center, launched his own war on poverty in the West End. His projects included opening a small loading pallet factory, which employed seven community people and projected a work force of thirty, a cooperative meat market, a thrift shop, and a furniture store. These activities won him friends in the West End, but his aggressive style made him enemies elsewhere. After he had been at DePorres five years, the archdiocese closed the center[11] and reassigned Sicking to St. Paul's Church in the Over the Rhine district, a poor and then predominantly white neighborhood which lay directly east of the upper West End and north of the central business district.

Tom Strattman, a Black Turk "general," claimed that the group formed to "keep order" after "our close friend" Father Sicking announced the closing of DePorres. The Church then moved activities associated with DePorres to the St. Ann Community Center in Queensgate II, renamed it the Martin Luther King Center, and appointed Strattman director. Strattman insisted the Turks wanted merely to achieve a measure of racial pride and community self-help and security through inculcating internal discipline and carrying out a variety of community improvement, clean-up, recreation, and development projects.[12] Many outsiders, however, took another view, in part because at least 7 other such groups existed in Cincinnati's black neighborhoods. Together, these organizations contained a membership estimated at 200 persons, almost all of them males between the ages of 16 and 25, some of them reportedly former members of Chicago's Blackstone Rangers. The Avondale Army, which operated in Cincinnati's largest black ghetto, ranked with the Turks as the area's biggest group, and it featured marching and the ceremonial use of machetes and "honor guards" as part of its community security patrol program. The police and press regarded this group as dangerous, though Kenneth Robinson, the Army's chief officer, stressed publicly that the group was "militant but not violent." Nonetheless, it and the Turks suffered from an "image" problem throughout their brief existence.[13]

During the summer of 1968 the university planning team sought to work with the Turks by contracting with the King chapter to gather environmental health data in the Center's clean-up campaign.[14] We badly needed the information and found cooperation with a resident group an attractive idea. Two black University of Cincinnati medical students wanted to join and direct the survey but several snags aborted the project, including the publicized arrest of one of these medical students in Chicago in the company of Black Turks participating in a motorcade for presidential candidate Eugene McCarthy.[15] Chicago police found four 24-inch machetes in the car, and placed the student on trial a week after the arrest. That incident seriously jeopardized the project, and Cincinnati's Department of Urban Development frowned on such resident contracts as well as on any seeming diversion from the main task, although activist members of the Task Force praised the proposed survey and similar efforts.

This incident, along with the riots, the 1969 campus disorders in Cincinnati, which closed the University of Cincinnati for a few weeks, and the activities of the Turks, Black Panthers, Blackstone Rangers, and Avondale Army, created a larger environment of militancy for the actors in the Queensgate II planning. On the other hand, a smaller and more intimate environment of militancy for resident participants and their spokespersons developed in small group discussions of social structure and the process of social change, at evening or weekend seminars in the homes of some of the Task Force participants in the West End. As one seminar member put it, "We would 'rap' until 2:00 in the morning over beer, wine and chicken, in the kitchen or living room."[16] I attended one such gathering in the home of a black member of the Task Force led by Robert Rhodes, a young Ph.D. student from the University of Chicago, who developed an analogy between European colonies and black ghettoes in America. "These sessions," in the words of one member of the seminar, "gave us activists an abstract, theoretical framework for community development and the economics of growth, and an understanding of the relationship between 'growth' and 'development.' We could look at parallels in Africa and Asia."

The local environment of militancy also included learning the methods for acting as catalytic "change agents" in community development work and

social and city planning. Paul Henry and Jerome Jenkins taught these techniques in the university's Graduate Department of Community Planning and
the training included confrontation in a tense atmosphere created by the use
of challenges, threats, epithets, invective, and issue raising, sometimes
subtly and sometimes brashly done. In May 1969, at a critical juncture in the
Queensgate II planning, Paul Henry and others gave Jay Chatterjee and me a
bitter taste of this technique as part of the activists' attempt to influence the
planning in favor of the residents.

We were invited to Harry Martin's home in Kennedy Heights ostensibly
for an evening of refreshments and discussion with Martin, Henry, Rankin,
Lewis, and Hugh Guest. We were all black, except Chatterjee who, though
dark, was from India. We assumed we were invited because we were among
those university planners who had come to feel a closeness to the community. I thought perhaps this was another of those informal seminars in social
structure and economic change. It was not.

It was a slow-developing, no-warning, nerve-wracking confrontation
session, with Chatterjee and me as the surprised "confrontees." In an apparently innocent and even humorous beginning, our hosts recounted their
childhood experiences of growing up in the West End and their memories of
other Cincinnati black neighborhoods, citing with affection certain stores,
vacant lots, intersections, hangouts, "marvels," pathways, and structures.
They also described how they had observed over the years a pattern of
"rip-offs," "takeovers," and physical "intrusions" by white chain stores, oil
companies, service stations, medical and educational institutions, and new,
suburban-oriented high-speed expressways, all "stealing," "destroying,"
"changing," and "exploiting" many of the sights and sites with which they
were long familiar, and all for the economic and practical benefit of the
outside, established, affluent white community. The pattern of exploiting
black neighborhoods had become predictable to them. Now they wanted to
confront us with an issue.

The pending crisis in Queensgate II planning at the time was a serious,
mounting dispute between the residents and the Department of Urban Development over certain policy directions for social and physical design. There
were signs that the university *administration* might side with City Hall. In a
threatening, challenging, deeply resonant voice, Paul Henry put the question to me: "Where do you stand, as university *planner*?" I responded about
how "we," the planning team, felt. He interrupted me brusquely and, looking menacingly straight into my eyes, said: "Not *'we'*; where do *you, Tom
Jenkins,* stand as *director* of the planning team?" I was nervous and tense,

but the point was clear. My apparent friendliness with the community and the fact that I was black cut no ice. If the chips were down, what would my performance be?

They then spelled out the larger meaning of their message. Though small (117 acres), the location and significance of Queensgate II was important for the overall inner-city black community in Cincinnati. It meant an immediate major investment of public and private money. Action there, with planning and construction, could boost community development morale in other black areas. The *kind* of action there could set a precedent for City Hall and the white establishment in dealing with other black neighborhoods. If necessary to protect the black neighborhood's interest in the project, Henry added, "we could bring black bodies from all over Cincinnati to physically squat or lay on the land in Queensgate II. You can tell all of those concerned they are not dealing with just Queensgate II!"

That evening is indelible in my memory. The experience was upsetting, for the charged atmosphere contrasted sharply with the tone of our previous relations. As we later learned, this was a conscious technique and ploy, but serious of purpose, and had little to do with personal relations, before then or since. The incident also underscored a minor irony of the neighborhood control dimension of the Queensgate II experience. Well-educated blacks seeking to advance the interests of the black Queensgate II neighborhood also spoke for the general interests of two larger "communities," the black population of the Cincinnati area and, given their interest in the colonial analogue for the plight of blacks in urban America, ghetto residents in big cities across the country.

Other attempts to exert resident control outside the Task Force framework also took place. In 1969, the West End Community Council halted land acquisition for Queensgate II by persuading the City Council to refer an acquisition proposal to committee until the resolution of the Task Force-Krabach argument over whether the Task Force had decision-making or advisory powers.[17] Similarly, Governor James Rhodes's proposal to build a $12 million two-year technical college in Queensgate II went down in part because West End residents joined members of the Task Force and others in angry newspaper protests against the scheme.[18] And when prospective developers sought to avoid Task Force review of their plans, the West End Community Council urged the designation of its Area Planning Committee as a "watchdog" over all development projects in Queensgate II and the entire West End so, as Doris Brown remarked, residents would not be left "in the dark about impending projects which will affect their lives."[19]

Resident participation inside and outside the Task Force proved more or less effective in all these circumstances, but residents and their closest allies in the West End and black communities did not stand alone in the effort to protect neighborhood interests against traditional planning practices and policies. The emergence of advocacy planning as an accepted practice among professional planners also played a part in citizen participation in the Queensgate II project. To reiterate Davidoff's argument: The advocate planner, like a lawyer, should represent the views of "have-not" clients against the notions of traditional planners who purported to serve the "public interest" but in reality protected the "haves" who exercised a disproportionate influence within the standard city planning structure.

Black members of the Task Force, of course, knew about advocacy planning, and in the summer of 1969 sought to secure an advocate planner for the Task Force so that they might better judge and evaluate the effect of planning, zoning, and relocation proposals made by the city or the university planning team. This issue emerged during the conflict with City Manager Krabach over the decision-making status of the Task Force, and it took the form of a demand to hire professional staff for the Task Force. That demand went unanswered, and ultimately the university planning team hired Mrs. Lucille Burden as neighborhood liaison out of funds from the city-university contract. Seven Hills Neighborhood Houses gave her an office in the Queensgate II area, and though she performed ably she lacked the professional training and experience to evaluate technical plans and hence fell far short of supplying the Task Force with an advocate planner.

But some advocacy planning appeared in irregular form on the Queensgate II project. The City Planning Commission Staff, with the assistance of personnel from the Department of Urban Development, prepared a 1966 report, "A Program for the West End Community," which recommended the creation of the Task Force with resident representation, and a year later the Planning Commission staff helped the Task Force draft the "Phase Two Policy Decisions," which, drawing on the 1966 report, bore directly on Queensgate II in a way which protected resident concerns.[20] Beyond that, Planning Commission staff performed administrative and clerical chores, took on survey and writing assignments for the Task Force, and helped make some early presentations to the members of the Planning Commission and the Department of Urban Development, all of which indirectly benefited residents.[21]

On several occasions, moreover, the university planning team adopted an advocate or quasi-advocate stance. It supported "adaptive planning" as an

approach to business relocation in order to better meet the social and economic condition of small and marginal resident businessmen. The team
recommended to the City Manager on behalf of the residents that the city
improve waste collection, sidewalk repair, weed control, and building and
fire inspection in Queensgate II *during* the planning process. The team
backed strongly the idea of promoting employment opportunities for residents in land use plans and development. It insisted upon a humane family
relocation plan to provide for one-move-only on-site moves, people-oriented staging, and the use of neighborhood organizations to facilitate citizen
participation in relocation. And the team supported the health survey advocated by the Black Turks and the university's Student Health Organization in
part as a means of demonstrating the residents' anxiety about rodent control
and eradication.[22]

However useful the university team and Planning Commission staff may
have been, their advocacy of residents' interests came from outside the Task
Force and constituted informal, ad hoc reactions to particular situations.
Neither they nor the hiring of Mrs. Burden compensated for the absence of a
full-time professional advocate planner paid for by the Task Force and
directly susceptible to resident influence. In that perspective, the Task Force
as a partner in the planning, and one of its factions, the neighborhood
residents, lacked the technical resources and therefore the power commanded by the city and the university planning team.

Other factors also inhibited the participatory role of residents in the
planning process. The Task Force held its regular meetings in St. Peter-in-
Chains Cathedral within the Queensgate II boundaries. But the late afternoon meeting time (3:30 to 5:30 p.m.) presented a problem for poor and
middle-class working people tied to the discipline of the clock, while conferring an advantage on business and professional types, including government
employees and private social agency staff, who either could afford to take
off from their work or who could count attendance at such meetings as a
legitimate part of their jobs.

Most Task Force meetings, though open to the public, reflected this bias
in favor of business and professional interests. Employees of the YWCA,
Shillito's department store, antipoverty agencies, Model Cities, the Cham-
ber of Commerce, city departments, and major churches attended most of
the open meetings, and, whether they sat on the Task Force or not, made
themselves heard through speeches and by buttonholing members. But
working-class residents of the neighborhood, small businessmen, public

housing occupants and renters, young people from the West End, and members of small West End churches seldom attended. Their views came before the Task Force only as filtered through the three resident members of the Task Force and their black and white allies chosen to represent other interests. Clearly, the residents had to rely on their advocates—Mallory, Jenkins, Lewis, Rankin, and the rest—to translate their desires into Task Force plans and actions.

In my view the Task Force recommendations through the planning stage clearly reflected the influence of the residents' representatives and their allies. But the Task Force's will was frustrated as the process moved from planning into the crucial development and design or implementation stages. Part of the failure stemmed from the Task Force's inability to create a monitoring system to see that developers and designers adhered strictly to the plan during implementation. The hiring of an advocate planner for the Task Force would have helped, as would a formal decision to act on the Task Force resolution June 1969 to retain the university planning team through the design and development stages. To be sure, the Task Force later secured the services of two Cincinnati architects sympathetic to the spirit of the plan, but their experience, too, proved frustrating.[23]

Equally important, however, was the loss of the high-powered, firebrand leadership and activist talent which proved so effective during the planning stage. Beginning about 1970 important blacks on the Task Force and their outside allies increasingly missed meetings, and concomitantly the Task Force, which met as often as twice weekly while engaged in planning, convened less frequently, occasionally as seldom as once every two months. The thin spread of civic leadership in a dying and dwindling neighborhood[24] explains some of the attrition, as does the natural turnover in positions inevitable in any project or institution engaged in a program of five years' duration. And the reasons for the black departures suggests the value of their participation, in view of their subsequent assignments with broader responsibilities (see Table 3.2). In any event, after 1970 development and design decisions altered significantly both Task Force recommendations and City Council policies.

The list of changes is long, and nowhere more critical than in the Town Center proposal. By 1973 revisions and proposed revisions substantially changed the concept hammered out by the Task Force. First, critics argued that an increase of commercial and office space adopted during the initial construction stages of the Town Center would adversely affect later rent

TABLE 3.2   Activities and Accomplishments of Key Citizen-Participants of Queensgate II Planning and Postplanning Periods

Citizen-Participant	Activities and Accomplishments		Notes
	Planning 1968-69	Postplanning 1970-74	
Doris Brown	Public school cafeteria kitchen helper; homeowner; community leader; President, Queensgate II Club; President, West End Community Council (WECC), West End Task Force (WETF)	Staff Director, Residential Neighborhood Community Assn. (RNCA), Citizens' Arm of Model Cities, with staff of 27, including a planner and beginning annual budget of $200,000	Natural leader, though little education; learned planning and related by experience; kept active schedule, full appointment book; trips to Washington, D.C. (re renewal funding); U.S. new towns; participant in national citizens' conferences
Jerome Jenkins	Professional social worker; Seven Hills Neighborhood Houses, Inc., Assistant Director; community organizer; WECC consultant; WETF; adjunct planning faculty, University of Cincinnati (UC); community studies and social planning; sensitivity group strategies	Graduate planning student, sociology minor (1969-74), Ph.D. (1974); Executive Director, Seven Hills; elected to board of Charterite party, which won City Council majority with Democratic coalition in 1971	Maintained both practical and intellectual interest in political action and social processes; industriously collected materials and data on West End communities and planning project, part of which became a starting point for Ph.D. dissertation; helped conduct and supervise social surveys of West End
Richard Lewis	Professional social worker; community organizer, Seven Hills Neighborhood Houses, Inc.; West End Special Services Project (WESSP); (antipoverty program); WETF; WECC; early community liaison to UC	Director, West End Development Corp. (WEDCO), arm of WECC; community development consultant	Helped direct or supervise West End community surveys; guest speaker in UC planning and sociology classes; learned planning almost from the ground up

James Rankin	Professional social worker; community organizer, Seven Hills Neighborhood Houses, Inc.; WECC; WETF	Elected Ohio State Representative, 69th district; had later key assignment to Finance Committee in legislature; Chairman of its Subcommittee on Health, Education, and Welfare	His sister-in-law was elected delegate to Democratic National Convention; ardent advocate for free-standing suburban-type houses in Queensgate II, "with white picket fences on all four sides"; helped conduct West End community and population studies
William Mallory	Ohio State Representative, 72nd District; WECC; WETF	Became (Democratic) majority whip in State Legislature lower house; spearheaded formation of community's WEDCO as one of two developers of Phase I of Queensgate II plans, i.e., as local WECC citizens' group working with a nationally known developer from Washington, D.C.	Always a power in WECC, whether in or out of office there; often urged and sometimes buttonholed UC planners to give full weight to community's viewpoint and welfare
Lucille Burden	Housewife; Secretary, Queensgate II Community Club; WECC; WESSP; community's staff liaison to UC planning team	Secretary, WEDCO; community observer at meetings to select principal developer for Queensgate II Phase I; same for meetings to select developer for QII, Block C	Learned community relations work, planning, and planning and renewal terminology by study and experience; traveled to Washington, D.C. (for QII renewal funding); to national citizens' conferences; to U.S. new towns
Harry Martin	Model Cities coordinator; formerly director of city's Concentrated Employment Services (CES)	Appointed as deputy in a federal agency in Washington, D.C.	Became well-known and respected in national network of Model Cities directors and staffs; tutored Hu Guest as his successor; accepted Washington, D.C. post at a time when Cincinnati Model Cities' future looked uncertain

(continued)

TABLE 3.2 Continued

| | Activities and Accomplishments | | |
Citizen-Participant	Planning 1968-69	Postplanning 1970-74	Notes
Charles Collins	A City Planning Commissioner; former City Councilman	Head of Cincinnati area office of federal Department of Housing and Urban Development (HUD), including FHA, responsible for mortgage insurance, special loans, and grants	Considered himself to be neighborhood boy who made good, able to serve home community; came up through the ranks
Paul Henry	Assistant Director, Community Relations Office, UC; adjunct professor, Graduate Community Planning, UC	Associate Vice-President for Community Relations, UC	His function as an effective bridge between elements within UC and between UC and community (especially black community) well recognized

levels and undermine the Center's ability as central focus and multiuse complex. Second, a change in the location of the pedestrian bridge over Central Parkway from the Center to Music Hall would permit affluent patrons of the Hall to avoid passing through the Center, thus denying them easy access to activities and facilities in the Center and simultaneously denying residents of the new Queensgate II ready entrance to Music Hall. This change, in effect, built a wall between the two structures as if to segregate one class of Cincinnatians from another. Third, the city's out-of-town urban design consultant proposed a vocational education program for the Center, a change which conflicted with the original idea of major commercial activity to produce revenue and enhance the complex's economic vitality and increase job opportunities for residents of the area. Finally, a change in future plans for the Liberty-Dalton section of the upper West End contradicted another basic concept on which the Town Center rested. The original plan contemplated major industrial expansion in Liberty-Dalton as a means of providing a broader diversity of jobs for West End residents and transportation improvements to connect Liberty-Dalton to the social services and commercial facilities in the Center and residential units in and around the Center. But the subsequent switch to light industry for Liberty-Dalton meant that few residents of Queensgate II or the West End would qualify for jobs in such high-technology firms. In all these ways, the notion of the Town Center as a central core which would connect Queensgate II and its people to other parts of the city's life and to employment fell victim to the deterioration of planning policies in the throes of implementation.

Yet to some the undesirable design and development alterations reached well beyond the Town Center. One such change increased the permissible height of buildings in the first residential construction. The Department of Urban Development also insisted upon the addition of a large public swimming pool in a 600-unit low- to moderate-income area. And the same department also isolated a downtown middle-income residential high rise on Garfield Place from the presumably less affluent Queensgate II of the future.

Clearly, the vision of Queensgate II as a mixed-income, racially integrated neighborhood dimmed as the black activists, through no fault of their own, faded from the scene after 1970. The Task Force, to be sure, resisted the subversion of its plan, but "the opposition," according to one close observer, "was mild . . . because the old fighters were all gone by then." In the final analysis, the residents and their spokespersons proved to be the critical element in the Queensgate II experience, for after they left, the planning partnership of the city, university, and Task Force almost surely let slip the chance to realize the dream.

NOTES

1. Miriam Geiser, "Planning with the People," *Journal of Housing* 25, 6 (1960): 13-17.

2. Anthony C. Downs, speech, annual meeting of the Better Housing League of Greater Cincinnati, 1973.

3. At that time the University of Cincinnati had both a graduate and an undergraduate planning department.

4. *Proceedings of the West End Task Force,* covering meetings in 1968 and 1969. Prepared by Hubert E. Guest, then of the City's Planning Commission staff, who served as the WETF's project director.

5. The Cincinnati *Enquirer,* circa May 20, 1969 ("Rep. Mallory Reveals West End Housing Plan") and the Cincinnati *Herald,* May 31, 1969 ("Rep. Mallory Announces Housing Corporation").

6. Cincinnati *Enquirer,* circa May 29, 1969; Cincinnati *Herald,* May 31, 1969.

7. Cincinnati *Enquirer,* circa May 29, 1969. See also Cincinnati *Enquirer,* September 5, 1969: "Group in Queensgate II Asks for Full Development Authority."

8. For a record of Taylor's presentation to the Task Force meeting at the local YWCA, and the ensuing question-and-answer discussion, see *WETF Minutes,* April 9, 1970. The discussion shows vigorous give and take, reservations on both sides, Taylor's economic pragmatism, and the political acumen of J. Jenkins and Mrs. Brown.

9. The report on this occupies the bulk of the *WETF Minutes* of July 16, 1970.

10. *Report of National Advisory Commission on Civil Disorders* (New York: Bantam Books, 1968), ch. 17, pp. 410-483. Compare with Vol. I and Vol. II of the Queensgate II Policy Plan.

11. James Adams, "DePorres Closed; Father Sicking is Reassigned," Cincinnati *Post & Times-Star,* September 7, 1968.

12. Ben Kaufman, "Black Berets March to Promote Self-Help: But Police are Wary," Cincinnati *Enquirer,* August 23, 1968. See also Adams's *Post & Times-Star* story of June 6, 1968: "What Turks Want: Pride in Race."

13. James Adams, "Black Turks Have Public Image Problem," Cincinnati *Post & Times-Star,* circa August 1968.

14. See Queensgate II Development Program Staff memorandum to Student Health Organization and Martin Luther King Community Service, dated September 9, 1968, which outlines items to be provided.

15. Cincinnati *Enquirer,* August 19, 1968: "Black Turks Released."

16. Telephone interview with Rick Lewis, February 1975.

17. See Cincinnati *Post & Times-Star,* July 25, 1969; and Cincinnati *Enquirer,* July 26, 1969.

18. Cincinnati *Enquirer* and Cincinnati *Post & Times-Star,* both February 4, 1969.

19. Cincinnati *Enquirer,* August 8, 1969. Supporting Mrs. Brown's request, Jerome Jenkins said, "We want to make sure the policies voted by the Task Force are implemented."

20. These policy decisions were approved by the Task Force, November 9, 1967. (West End Task Force, *Phase II Policy Decisions,* December 1967.)

21. Mim Geiser gave the most active assistance from the Planning Commission staff, and helped write the report, *A Portrait of Health Conditions in the West End* (June 1968).

22. See *Queensgate II Plan, Vol. III: Background and Research Papers*.

23. David Lee Smith and Donald Stevens.

24. West End Special Services Project of Seven Hills Neighborhood Houses, Inc., *1968 Supplementary Survey for the Queensgate II Urban Renewal Project Area* (August 21, 1968), pp. 2, 2A, 3, 6. Also see WEESP's *1967 Summer Employment Survey,* and W. D. Heisel's 1969 report restructuring WESSP's population data.

CHAPTER 4

Queensgate II and "the Movement":

A VIEW FROM THE COMMUNITY

Jerome R. Jenkins
Richard W. Lewis

In retrospect we are struck by the view of the future taken by the residents of the Queensgate II community. They wanted the kind of total community development necessary to promote a fundamental change in the social class of blacks in Cincinnati's West End. And they believed that this view stood diametrically opposed to that of city officials who, the residents felt, equated urban renewal with massive black removal. The community was determined to prevent the repetition of "old style" physical urban renewal in Queensgate II, and that determination was reinforced by adherence to an emerging new philosophy stemming from the convergence of the civil rights and black power movements. Black people in the West End during the late 1960s adopted the new style, rhetoric, and ideology of "black participatory power," which reminded older residents of the local manifestations of Marcus Garvey's United Negro Improvement Association in the 1920s.

Looking back, it strikes us that the mood of the West End represented a logical, historical development characteristic of all revolutions. Such movements involve ever deeper layers of oppressed masses whose grievances are deeply rooted in the nature of the system and who are ready to take increasingly desperate actions against it. The black movement, here as elsewhere in the 1960s, began with a reform stage led by Dr. Martin Luther King. It sought civil rights and integration, and its strategy was confrontation through disciplined demonstration. This strategy exposed both the pitiful inadequacy of the concessions it exacted and the inhumanity of whites with whom the demonstrators hoped to integrate. The white response to King's *means* of confrontation taught black people that all the civil rights legislation in the world could not solve their real grievances. In light of white reactions to Dr. King's movement, blacks began to ask whether whites were good enough to integrate with. "Why," as the saying goes, "fight to get into a burning house? Why integrate with cancer?"

Dr. King, of course, never drew this conclusion from his movement. That came from young blacks, many of them members of SNCC who pursued King's strategy in the South. By the mid-1960s they had arrived at a position close to that Malcolm X preached to black audiences in the north after his break with the Muslims. Malcolm rejected Elijah Muhammed's notion of black power, which stemmed from the religious prophecy of an apocalyptic destruction of white power. Malcolm contended, instead, that black power could be realized only out of a struggle for power through black revolution. But his assassination prevented him from developing an organizational form for his revolutionary perspective.

The civil rights movement came North after the Birmingham riots of 1963, at just the time when young blacks were reformulating King's conception of the movement, and between 1963 and 1968 the black masses erupted in a series of spontaneous rebellions in virtually every major northern city. The Detroit rebellion in the summer of 1967 established black power as a national phenomenon and laid the basis for the stage in which blacks took "Black Power" as a slogan and began exploiting the fears and panic of the white power structure to gain certain kinds of power for themselves. Indeed, after 1967 black power became so respectable that practically everybody in black communities across the country became a black power spokesman. Some reduced the idea to its lowest common denominator and talked and practiced it as a cultural phenomenon, defining black power as black consciousness, black pride, and black identity symbolized by wearing Afro hairdress and clothing. But other groups within the black community began to put content into the slogan which best expressed their particular grievances, aspirations, and capacities. The Queensgate II community was not exempt from this process.

We and many of the personnel of the Seven Hills Neighborhood Houses had been active in SNCC, CORE, or the NAACP, the agencies most responsible for the black protest movement in the South and in Cincinnati. Our experience convinced us that any activity designed to bring justice to the black community must secure full citizenship for blacks and full access to the goods and services of American society, and must provide for the continuous development of black communities over time. We were determined that the Queensgate II program would not become merely another "black removal program."

Our experience also provided us with insights and skills and an ideology for successful interaction with the "stewards" of the system. We knew that the stewards in any city or country became reactionary when their basic values were confronted by demands for structural changes, especially when those changes threatened their social position and control. We knew, too, that the stewards controlled unlimited resources for squelching challengers. But we also knew that the stewards harbored a gut-level fear of blacks when they appeared to be unified behind a cause.

In addition, we recognized that Cincinnati was a city which domesticated blacks. It did not like blacks who became "uppity," and it tried to crush or buy off those who acted belligerently. Cincinnati's stewards, however, preferred to handle "uppity" blacks through smooth, back room negotiations, and as a result the strategy of "eyeball-to-eyeball" public confrontation was amazingly effective. The more "uppity" in public you became, particularly if you had a specific and well-thought-out program, the more progress you made. But confrontation, we believed, also had to be well timed to be effective, so that before the system adjusted to public "uppityness," you could present a program and rally community support around it. And since Cincinnati's stewards prided themselves on being an intellectual community, any program for blacks had to be well thought-out and clearly elaborated.

On this theory of social confrontation and change we set out to redress the grievances of blacks in the West End, an effort which, under the aegis of urban renewal, we expected to emerge as a program for black development over time. To frustrate the efforts of the stewards to abort our efforts through tactics of divide and conquer, we established several guidelines to govern our behavior through the campaign. We agreed never to accept "amenities" from the stewards, never to attend as individuals any meeting called by the stewards, never to confront the system without first completing our "homework," and, when at a meeting, to designate in advance one person to present our program and answer all questions, another to present our philosophical position, and a third to serve as the unrelenting advocate of our position. We rehearsed these guidelines and tactics repeatedly at strategy sessions in our homes. We were, in short, committed to heed the advice laid down by Dr. King at the Freedomway Memorial meeting for Dr. W. E. B. DuBois two months before King's assassination. "DuBois, above all," Dr. King noted,

> did not content himself with hurling invective for emotional relief and then to retire to smug passive satisfaction. History had taught him it is not enough for a people to be angry. The supreme task is to organize and unite people so that their anger becomes a transforming force.

We also recognized, however, that our efforts would be futile if we failed to strike an alliance with whites whose interests coincided with ours. Long ago, Frederick Douglass answered those who persist in seeing the destiny of oppressed black people as separate and unrelated to the destiny of exploited whites. "We deem it a settled point," Douglass wrote, "that the destiny of the colored man is bound up with the white people of this country . . . and the question ought to be . . . what principle should dictate policy." For Douglass, as for us, unity among oppressed blacks as the starting point toward the achievement of black power was fully consistent with unity with oppressed whites and coalitions with other strata of white society to advance the cause of liberation. We knew, too, that we must seek political equality as well as economic and social improvement for blacks, and it seemed likely to us in the mid-1960s that if our needs and aspirations were not met the city would face the prophecy of Langston Hughes, who wrote of "dreams deferred."[1]

This was our mood, strategy, and philosophy as the events which led to the formation of the West End Task Force and the development of the city, university, and community partnership and the Queensgate II plans unfolded. For us, it began on June 22, 1964, with a report from the city's Traffic Department recommending the demolition of two residential blocks south of Crosley Field, the home of the Cincinnati Reds, to provide more parking for baseball fans. That news stunned the West End Community Council, for its Area Planning Committee had been meeting with the City Planning Commission to develop plans for the renewal of the West End, yet knew nothing of the proposal to tear down houses to make room for cars. This event proved to be the straw that broke the camel's back, and it led to several important steps.

We decided first to respond aggressively to the Traffic Department report. With the help of Seven Hills's federally funded Special Services staff, four school principals, and volunteers, the Community Council prepared a study of its own which attacked the Traffic Department report as erroneous and devoid of consideration for the welfare of the residents of the area. Then in April of 1965 the Community Council issued a list of demands concerning the future development of the West End, including one that urged that the dormant Queensgate II project be recertified and pushed as a federally funded urban redevelopment program, another that called for the creation of a task force composed of representatives from the city administration and residents to work with the staff of the City Planning Commission to complete the planning for the West End and to carry out plans already prepared by the Community Council's Area Planning Committee, and a third calling on City Council to meet with residents to discuss citizen concerns for the West End.

Meanwhile, we moved to tighten up the organization of the Queensgate II neighborhood. In August of 1965 an organizer hired by Seven Hills Neighborhood Houses put together a new organization called the Queensgate II Community Club. It languished for a time, suffering from the lack of a permanent meeting place and continuity in its leadership. But after James Rankin took up the organizing responsibilities, the few remaining members visited homes in a recruiting campaign and surveyed the area for a headquarters site. They finally settled on a John Street location which formerly housed the barber shop of Henry Johnson, who, for a quarter of a century, functioned as the sage and confidant of the community in weekly meetings in the shop held for the discussion of personal and community problems. Acquiring that landmark identified the Club more closely with the neighborhood, and soon residents dropped in to see what was going on in the new "office." Thereafter the Club's membership began to grow, attracting 13 new members in its first month at the John Street site.

At the same time, during the late summer and fall of 1965, City Council and the West End Community Council held a series of meetings at Sands Elementary School in the West End. Attendance averaged over 200 at these sessions, and during each session irate West Enders peppered council members with hot and bitter questions about the entire renewal-removal process. The meetings climaxed with a walk-in demonstration at a regular meeting of City Council in city hall. Over 300 West End residents participated, and several interrupted the Council's deliberations with requests from the audience to be heard. Council refused, arguing that the demonstration amounted to intimidation. Tension mounted when detectives and uniformed police officers filed into council chambers to restore order. After things quieted down, Council promised the West End residents that a proposal concerning renewal in the West End would be forthcoming.

Before action occurred on that promise, however, another crisis emerged. In January of 1966 the city engaged in negotiations to secure a National Football League franchise amid rumors that the West End constituted the prime site for a stadium to house the team. The West End Community Council, of course, opposed the idea, and organized six groups to fight the proposal. The highlight of that effort was a press release and letter to council members put out by the Area Planning Council which laid out its objections and aired long-standing grievances of West End residents.

> Within the West End Community, there are . . . many factors which force us to take this stand: (1) The City promised a program for rehabilitation and code enforcement and no fulfillment of this program is apparent. (2) We asked for plans and we have received proposed zoning changes. (3) There has been consistent tearing down of substandard housing and there has been no standard housing planned or built. (4) Health facilities are desperately needed and money was allocated for a prolonged study. . . . (5) We have lost three schools and gained none. (6) Our recreation facilities have been almost eliminated and there have been no satisfactory additions. (7) Too rapid, unplanned displacement of people has resulted in their moving into substandard housing. (8) The promise of 15,000 jobs for Queensgate I failed to materialize. (9) The West End has contributed much to the progress of our City. Our people's homes have been sacrificed for industry and expressways. (10) Urban renewal has done nothing for the people of our community.

The release ended with a warning about the "raising of people's expectations without fulfillment of even a portion of the promises made" and an expression of concern over "the growing loss of faith by people in our City government and the increased restlessness being displayed as a result of not being included in planning for its people." "We expect planning *for people*," the release concluded, "and speedy implementation of such plans. Anything less makes a mockery of city government."

In February of 1966 the city acted. In a series of meetings moderated by the Cincinnati Human Relations Committee, the West End Community Council's Area Planning Council, several city council members, representatives of the city administration, and the City Manager laid the groundwork for the creation of a task force to plan for the future development of the West End. By the end of April 1966, the proposal had passed through the

bureaucracy, letters of authorization to organizations represented on
the Task Force had gone out, and the long process of finding a chair-
person acceptable to all parties began.

The early meetings of the Task Force were characterized by suspicions,
doubts, skepticism, and downright distrust on the part of West End residents. Memories of the Crosley Field parking fight were fresh in their
minds, and the city's past indifference to West End conditions formed a
bedrock of cynicism about the city's good intentions. At the very first meeting, moreover, a West End minister who had not participated in recent
Community Council affairs questioned the eligibility of the West End dele-
gates on the Task Force. Several of the West End representatives, in addition, questioned the advisory role of the Task Force, fearing that it lacked the
power to produce real results, and some viewed the Task Force as a stalking
horse to divert the attention of the West Enders from the real plans of the city.
Despite these difficulties, however, and after months of difficult study,
rough-and-tumble debate, hard questioning, and working together, the Task
Force gained credibility and jelled as a policy-making body that could
produce firm planning decisions. It was, in the words of the Reverend
Robert Beck, a West End Community Council representative on the Task
Force, "a most unique experiment in democracy." "Frequently," he later
remarked,

> we have deferred planning policy decisions so that we could return to our
> respective groups, bring them up-to-date on Task Force activities and the
> decisions which have to be made, and to obtain their reactions and sugges-
> tions. . . . While I'm not out of sympathy with the tediousness, slowness, and
> frustration that necessarily evolves from a procedure which requires such total
> community awareness and involvement, nonetheless I'm aware that we are
> operating within the very purest kind of democratic framework.

It was a process, he argued, "practically unparalleled in local legislative
history."[2]

As the Task Force struggled to get off the ground, we continued to build
the Queensgate II Community Club. On March 20, 1966, we held an open
house and ribbon cutting ceremonies for the official opening of the office.
As more members came in, the group became more active. It focused
attention on the bad conditions at the Ritz Skating Rink, fought for a traffic
light at the intersection of John and Elizabeth Streets, and joined with a
group of University of Cincinnati fraternity members to clean up and refur-

bish a church as a recreation center. Slowly the idea caught on among the small residue of middle-class residents—homeowners, small businessmen, ministers, retired school teachers and postmen—who still lived in the area, and the Club flourished and gained confidence. The Club made its needs and desires known to the West End Community Council, the West End Task Force, City Council, and members of the city administration. Its proudest moment came in May of 1968, when members of the Club joined city officials in a trip to Washington to pressure the Department of Housing and Urban Development for action on the city's grant proposal. The Club, however, paid its members' way by holding barbecues, fish fries, and bake sales to raise money.

All these efforts constituted tactics in our general strategy for shaping the Queensgate II program. We saw the objectives of the urban renewal project as the promotion of the social, economic, and cultural development of the community. We had to educate and secure a commitment from city officials to look upon the project as a *process* which emphasized that the residents must change socially and psychologically; as a *method* which emphasized the ends to be achieved; as a *program* with specific elements to improve health, education, employment, housing, and recreation; and, finally, as a *movement* which emphasized an ideology of individual worth and dignity. We believed that the two essential elements for the success of this kind of Queensgate II project would be participation by the residents themselves in efforts to improve their level of living, with as much reliance as possible on their own initiative, and the provision of technical and other services to the residents in ways to encourage initiative, self-help, and mutual help. These techniques, we felt, could be applied to promote specific results that would develop into a movement with an emotional and ideological appeal that other black communities could utilize as a model for change.

At the least, we hoped to recreate the spirit of community and diversity which once characterized the West End, and memories of the neighborhood in the 1920s and 1930s fired the determination of the residents to improve its future prospects. Though segregated from the mainstream of Cincinnati's life, the black people of the West End had developed by the second quarter of the twentieth century a spiritual and cultural identity expressed in a variety of institutions. Black churches flourished, and the list included four Baptist congregations, five Methodist churches, two Catholic parishes, and one Episcopal parish, as well as a host of storefront groups. The area also contained leisure facilities, among them four movie houses—the Roosevelt,

Lincoln, Dixie, and Pekin—and two famous night clubs, the Cotton Club and another in the grand old Hotel Sterling. Social and political organizations included Democratic and Republican clubs in the Sixteenth Ward, the West End Civic League, the Elks and Masonic lodges, and the Universal Negro Improvement Association. The Black 372nd National Guard Company also helped foster unity and pride, and youngsters had a counterpart in Boy Scout Troop 55.

The Men's Lodge of the Ninth Street YMCA played an especially important role. It functioned as the West End's "City Club," a place where black leaders in the West End met to discuss the affairs of the day. Participants included Jesse D. Locker, a politician and Republican city councilman, Judge William Lovelace, the city's first black judge, Mr. Phillis, principal of Stowe School, Drs. Clark and Paine, Attorney Posey, the Reverend Jesse Chiles, Myron Bush, Sr., a lawyer, Myron Bush, Jr., who became a city councilman in the 1960s, and Theodore M. Berry, who became a city councilman and ultimately mayor of the city in the early 1970s. The West End also had a less formal "City Club" in the form of soap-box oratory presented in Sinton Park on the corner of Kenyon and Mound. No topic was proscribed, and many a youngster acquired a liberal education just by standing and listening to the exchanges between speakers and the crowd.

In short, a spirit of community existed in the "old" West End of the early twentieth century. The place possessed a small-town atmosphere, and the link of person to person and group to group was close. With the dispersal of West Enders after urban renewal, however, the feeling of togetherness in the West End disappeared. We hoped to retrieve and revivify that feeling, and one of our first steps in the process was to develop a plan for a viable community in Queensgate II for the latter half of the twentieth century.

There remained a critical obstacle to surmount before the planning process could begin, however. The residents recognized that they lacked legal, financial, land use, and architectural expertise for community planning. City Hall was one repository of such expertise, but it was not disposed to provide this expertise to the community on a continuing basis. Besides, the community knew that the deterioration of the community was the direct result of the city's urban renewal program and feared that in Queensgate II, as in other places, the city would remove blacks to a new ghetto and turn the old area into a showcase downtown development of commercial office buildings or high-rise apartments for young liberated urban types who wanted to be swingers. This sort of renewal aimed at renovating an urban area, instead of attacking directly the real problems of black unemployment, segregated housing, poor health services, inadequate transportation, and quality city services. In short, the city's traditional promise of decent housing elsewhere would not suffice. We believed a new breed of planner and planning process must be introduced into the Queensgate II community if it were to survive and develop over time.

A crisis in race relations during 1966 and 1967 lowered the credibility of the city among blacks and intensified our conviction that Queensgate II required a break from conventional planning techniques and processes. Since 1948, when the city's master plan projected central-city redevelopment through a program of minority group removal and relocation, each slum clearance project uprooted another segment of the black population. The black population, in effect, paid the social costs of remaking the central business district. Never was there a more enthusiastic pursuit of the doctrine that whatever is good for business is good for Cincinnati. People who opposed relocation for urban slum clearance were labeled obstructors of public progress. This attitude was in the best interests of the city's giant corporations, and the program emerged as a consequence of an alliance between the rich and powerful and the city. It built on systematic violence, project by project, highway by highway, home by home. A great city was being rebuilt, but a defenseless people was also being ground into the rubble. Mounting dissatisfaction with this alliance of City Hall and the stewards, combined with their plans for blacks, combined with insults produced the rebellion of 1967.

It is difficult to state precisely which insult triggered the violence of the 1967 Cincinnati rebellion. But some feel the arrest and sentencing to death of Posteal Laskey as the "Cincinnati Strangler" ignited the outburst. Many in the black community believed that the police were hard pressed by a fear-ridden city to find and convict the individual responsible for a series of strangling murders committed during 1966. Many also felt that Laskey, a local black jazz musician, received an unfair trial. A conviction was needed and it was, they concluded, no surprise that the man convicted turned out to be a black. To make matters worse, a white newspaper copyman about the same time got off with a suspended sentence when convicted of manslaughter for the murder of his girlfriend. The city seemed to be following a double standard of justice, one for whites and one for blacks.

Tension increased in June as a result of two other developments. The Laskey case occurred in the context of a stringent effort to enforce the city's antiloitering ordinance, a campaign which netted 240 loiterers, 170 of them black, between January and June. And that month Laskey's cousin, Peter Frakes, was arrested for allegedly obstructing pedestrian traffic while carrying a signboard which read "Cincinnati Guilty—Laskey Innocent."

On Sunday, June 11, 1967, a Negro mass meeting convened in Avondale, the city's major and newest ghetto, to discuss recent events in the light of accumulating black grievances. At another rally on the night of June 12 the discussion centered on the pros and cons of police enforcement of the antiloitering ordinance. After this rally, young blacks firebombed a drugstore on Reading Road and broke windows in about two dozen stores while fighting erupted between blacks and whites. The police arrested 14 and the judge announced in advance of the trial his intention to mete out maximum sentences to anyone convicted of a riot-connected offense. Yet most whites were booked and sentenced for disorderly conduct, with a maximum fine of $100 and 30 days in jail, while most blacks stood trial for violation of the riot act, an offense which carried a maximum fine of $500 and a 1-year jail sentence. Militant leaders demanded amnesty, release of the prisoners, and guarantees of equal justice, but City Council refused to meet with the militants at a site in Avondale. On the following evening "all hell broke loose" and the city entered its long, hot summer of 1967.

The stewards failed to respond to any of these events in a constructive way. Traditionally, decisions about the welfare of the black community were made by top-echelon business leaders after negotiations with "responsible" middle-class Negro leadership. The vehicle for these negotiations in the mid-1960s was the biracial Committee of Twenty-Eight, which, early in 1967, had announced as its goal the provision of 2000 jobs for young Negroes. At the peak of the campaign, 65 jobs had been produced.

But this "solution" of the problem was much less important than the manner in which it was accomplished. Establishment whites and Negroes were parties to the talks, and black militants were specifically excluded. The solution was, in other words, a gift of grace, not a response to a demand. As a gift of grace, it was clearly understood that the gift could be granted or withheld at the pleasure of the two establishments. In the aftermath of the rebellion of 1967 the Committee of Twenty-Eight adhered to its basic notion that the needs of blacks in the area of employment and economic justice were subject to the good pleasure, if not the whim, of the two establishments.

The abortive attempt of middle-management business leaders in cooperation with the Cincinnati Industrial Mission to form a Cincinnati Urban Coalition which would bring together the stewards from the big corporations with militant black leaders serves as a case in point. Week after week the middle-management people met with the black militant leadership, but when the top level executives were asked to lend the movement their encouragement and moral support they remained silent. When pressed, they complained that the black leadership included people like Frakes and another black who had been convicted of carrying a concealed weapon. The stewards had no stomach for dealing with militants, and they made this known to their junior executives and middle-management subordinates. Clearly, they viewed the rebellion of 1967 as a threat to the traditional forms of business leadership, and as that fear became obsessive law and order became a stronger consideration in decision making than problem solving. As time went on it became clear that the Committee of Twenty-Eight, the business community, and the Republican Party were taking on the role of defenders of the public peace. This turn of events, in our view, formed the backdrop for the appointment of Richard Krabach as City Manager, and for the rising law and order mood among the city's leaders which pushed the credibility of City Hall to an irreducible minimum. It now seemed more imperative than ever that the planning process for Queensgate II be handled in a way that would maximize the role of residents, compensate them for their lack of planning expertise, and minimize the role of city agencies and stewards tainted in the minds of blacks with a generation of neglect and misuse.

Opinions differ as to who initiated the idea of bringing in the University of Cincinnati as the planning consultant, but to the best of our knowledge we first raised it shortly after the establishment of the West End Task Force. We knew several professors at the university who seemed sympathetic to our struggle, and City Manager William Wichman and John U. Allen, then head of the city's Department of Urban Development, agreed with us that both Urban Development and the City Planning Commission suffered from low credibility among blacks. The city had, moreover, used a citizen's panel working with outside professional consultants in developing plans for the revitalization of downtown, and the leadership then in City Hall believed that worked well. During the fall of 1967 Wichman raised the issue of university participation with a university vice president who, in a month or so, indicated that the university was enthusiastic about the idea. We wanted a black to direct the project, and after the university came up with Tom Jenkins from Boston and other parties concerned agreed to the tripartite arrangement, we made the announcement to the press in May of 1968. As we understood the partnership, the planning activities of the university were to be in behalf of the residents of the Queensgate II community and the community possessed the veto power before proposals went to the City Planning Commission and City Council for final ratification. The organization charged with the responsibility of approving the plans developed by the university, then, was the West End Task Force, and the Task Force became a bridge between the parties with an interest in the neighborhood, the city government, and the university planning team.

The entire planning process involved multidimensional conflicts, as subsequent interviews with participants revealed. One important figure in City Hall, for example, claimed that both the Department of Urban Development and the City Council felt apprehensive about bringing in the university because of the impractical, "egghead" image of university faculty, and that Urban Development often had to remind the university planning team to meet deadlines. Peter Kory, the Assistant Director of Urban Development, who took over as Director midway in the process, sometimes thought the university was moving too slowly and going off on the wrong track, and he and the Director of the Department had to work out some differences between themselves over goals. Yet both agreed that they should "guide" the university without creating the impression that they were telling the university what to come up with. Wichman's successor as City Manager, Richard Krabach, lacked confidence in the participation process the city worked out for both downtown and Queensgate II and wanted the city or "professionals," not citizens, to make the decisions. He grew indifferent and failed to attend Task Force meetings, and one key person in the Department of Urban Development resigned in part because of the absence of support from the City Manager. After resigning he admitted to a lot of impatience in City Hall with the "groping" that went on and with the "education process" the city had to go through with the West End Task Force and the university. He recognized, however, that no other university had done this kind of thing before and, despite all the difficulties, remained optimistic about getting a good plan out of the arrangement. He came away feeling that a good job was done, and claimed that the only way to "produce in the urban renewal business" was to "look at the past, and say maybe that was good at the time, and maybe those decisions were sound at the time, but in light of today's thinking those directions were bad. So let's see if we can correct them."

An important member of the university team, a former city employee, also felt distress at the "groping," and ran into conflict situations. He objected to having a history of the neighborhood done on the grounds that the plan could have been done without it and that the expenditure of contract funds for that sort of thing was never really thought out. He felt that the university's central administration feared from the outset that the process would not work out and that the university would be left holding the bag. He also reported that some administrators, in the middle of the work, questioned the wisdom of undertaking such a tough problem when some of the principal actors in City Hall and on the university team could not seem to get along with one another, and he was at times called in to act as a referee between one figure from City Hall and two members of the university team who more than once engaged in table-thumping shouting matches. In addition, he felt the top leaders of the university team wanted, in effect, to "write a book," while Urban Development more realistically saw the project "as an action planning type of thing which should result in a product that was an ordinance . . . that would set the goals for the area, which was finally done." He argued with the team that writing a book would permit the distortion of the plan by those who prepared the final ordinance. He believed, too, that the most controversial figure in City Hall, who once antagonized the team by trying to move the project's headquarters into City Hall, would not have been given credit for making an honest mistake but would have been seen as "malicious" by the university team leadership. He recalled, too, that university vice presidents proved sensitive to criticism of the university in the press or by city council members and brought those problems to the team. But he contended that all of the problems got ironed out and the partners produced a good plan.

Residents of Queensgate II, of course, distrusted City Hall from the outset, but we were more sympathetic toward the university because we regarded it as a potential planning advocate for the Queensgate II community, as well as a source of expertise. We never, of course, fully trusted the university. We suspected, and later our suspicions were partially confirmed, that the team was under pressure from downtown interests and from the Department of Urban Development to follow their lead. Indeed, at one point the president of the Queensgate II Club invited the president of the University to appear before the Club because many residents felt City Hall was controlling the university team. And at another time some black members of the West End Task Force asked the team's director and associate director to the home of one of their friends for what the director called "some confrontation and sensitivity training," in an effort to find out "whether we were leaning with the black members of the Task Force or City Hall."

The legacy of distrust produced by over twenty years' experience with traditional planning procedures in Cincinnati, our uncertainties about the university, and our fear that the plan might be gutted in the process of carrying it out, provided part of the motivation for establishing the West End Development Corporation (WEDCO). This was an organization designed to implement the plans and thereby enable us to watchdog the entire process of completing the project. State Representative William L. Mallory played the key role in organizing WEDCO, but Councilman and Mayor-Elect Theodore M. Berry and Peter J. Randolph, a black attorney, helped give structure to the idea. The Cincinnati Human Relations Commission also played a part in the endeavor. It began to take shape early in 1969 when the leadership of the Queensgate Community Club and several members of the West End Community Council held a series of meetings to create a nonprofit housing development corporation. After months of meeting, the state, on May 13, 1969, chartered WEDCO with a mandate to provide housing and related facilities for employment and services to low- and moderate-income families.[3] It was a black corporation whose board of directors was composed of a minister, a social worker, four community workers, a postal worker, a barber, a beautician, a cook, a schoolteacher, and a coordinator of an apprenticeship minority outreach program.

Although WEDCO won the respect of the city administration and could call on a wide range of local expertise for aid, we knew it lacked sufficient muscle to develop a major renewal project. WEDCO therefore launched a search for a developer to assist it in competing for the contract to develop Queensgate II. We heard about Mid-City Developers of Washington, D.C., and Mallory and the WEDCO board visited the capital to meet principals in the firm and view some of the work completed by Mid-City in Washington. We decided Mid-City was a progressive enterprise with a philosophy of community development and control that suited the philosophy of the residents of the West End. WEDCO and Mid-City signed an agreement in August of 1970, and on September 15, 1971, WEDCO and Mid-City developers and the city of Cincinnati signed a contract for the development of the first phase of the Queensgate II project, a project covering an area bounded by Ninth Street on the south, Court Street on the north, Central Avenue on the east, and John Street on the west.

Despite our successes, we were, in the final analysis, disappointed with our experience in partnership planning. Much of our disappointment centers on the university, for it seemed our only potential ally against the stewards and their friends in City Hall. Our experience suggests that the university possessed the academic and technical expertise to help the city solve urban problems, but it lacked the political savvy to translate its expertise into functional programs. On the other side, the city lacked the flexibility to accept the kind of prescriptions the university could provide unless the prescriptions coincided with the views of the prevailing political and economic power structure, which, not insignificantly, could exert influence on the university through its politically appointed board of directors. The university faculty, moreover, was subject to the social, psychological, and economic pressures and biases which afflicted the general society, found it difficult to accept the value of the social structure and norms of a low-income black community, and hence ranked as part of the problem. And collaboration between the university and the city inevitably produced conflict because the municipal bureaucrats viewed themselves as "practical men" who could get things accomplished, and perceived the university as a place where "egghead" faculty spent an excessive amount of time thinking and formulating theories. In short, and quite apart from the fact that the complexities of urban planning in Queensgate II were designed to include public involvement and "checks and balances" on planning officials and their programs, genuine citizen participation was not accepted by institutions with political or economic power because the citizens' views seemed detrimental to the power base of the controlling institutions.

Out of the Queensgate II experience comes three recommendations. Those who participated should synthesize and analyze the process, but if that evaluation produces only criticisms the consequence will be despair, negativism, frustration, and the discouragement of future involvement by universities. Those who failed to participate in Queensgate II should ignore the critics who say, "this and that should have been done," and "I told you so." To those seers we say there are many urban problems that are unsolved and which beckon to your theories and expertise. Yours is only to have the initiative and courage to get involved and meet the challenge. Finally, to black residents of urban communities like Queensgate II, we recommend to you the words of the late Tom Mboya, a leader in Kenya's drive for independence, as we face the realities of the future:

> We suffered during our struggle for freedom, but in many ways it was a simpler period than today. It was one of mass mobilization, dramatic demonstrations, and profound nationalist emotions. The present period is less dramatic. Fewer headlines are being made, fewer heroes are emerging. National-

ist sentiment must remain powerful, but it can no longer be sustained by slogans. . . . Rather, it must itself sustain the population during the long process of development. For development will not come immediately. It is a process that requires time, planning, sacrifice and hard work.[4]

NOTES

1. Langston Hughes, "Lennox Avenue Mural, I: Harlem," in Arna Bontemps, ed., *American Negro Poetry* (New York: Hill & Wang, 1963), pp. 67-68.

2. Richard W. Lewis, "Planning for Change." West End Community Council Working Paper, circa 1969, p. 1. (unpublished)

3. Richard W. Lewis, "Coordinated Approach to Planning: WEDCO." Urban Studies Collection, Archival Collections of the University of Cincinnati, April 24, 1972, pp. 2, 16. (typescript)

4. See Tom Mboya, "Key Questions for Awakening Africa," in H. W. Rudman and I. Rosenthal, eds., *A Contemporary Reader* (New York: Ronald, 1961), pp. 358-363; see also Tom Mboya, "African Freedom," in Langston Hughes, ed., *An African Treasury* (New York: Crown, 1960), pp. 30-35.

CHAPTER 5

The West End Task Force:

THE VIEW FROM THE CHAIR

Edgar J. Mack

Queensgate II was my first experience with the politics of city planning. I had been active in securing support for the Cincinnati Symphony Orchestra, and I worked for the Community Chest and served as president of a home for the aged. But in the mid-1960s I knew virtually no one in City Hall, and only Charles P. Taft and Willis Gradison, Jr., on City Council. Though I had, years ago, used my influence as an executive with the Red Top Brewing Company to integrate the firm's work force, I had not been actively working with the black community, or any other community, for that matter. Red Top happened to be located in the West End. My only other connection to the area was even more remote. My great-great-grandfather used to live there, but in the 1960s I lived in Indian Hill, a silkstocking suburb north and east of the city.

To the best of my knowledge I became chairman of the Task Force because no one else would take it. I learned later that my wife's sister, who worked for a social agency headed by a black man, suggested me through her boss to the people seeking someone to chair the Task Force. In March of

1966, Bill Mallory, a black resident of the West End, asked me if I would serve as chairman. I asked what the West End Task Force was. He said it was a group of citizens from the West End raising hell with City Council for tearing down houses near Crosley Field to provide parking space so the Reds would not leave town and to make room for commercial and industrial expansion in the area. Mallory admitted that four or five other people refused to take the Task Force job because they felt council members merely wanted to get the blacks off *their* backs. I thought about the offer for three days. It seemed both an exciting challenge and a chance to help some people, so I called Mallory and accepted. He said he would check me out with City Manager William Wichman and call me back.

About a week later Mallory called and said the manager believed Council would not accept me. I asked why. He replied that Republicans dominated Council and that my voting record showed I had voted for too many Democrats. I blew up, but agreed to an interview with Wichman and four or five of his department heads and staff. They asked me a few questions, including whether I had any financial interest in the West End. I said no, and that was that. I felt like a damned fool for walking all the way from Fourth and Walnut to City Hall to answer half a dozen silly questions.

I did not hear from Wichman, so I finally called him. He said he was going to Washington that day and promised to see me when he returned. I happened to be going to Washington that day, too, and offered to talk with him on the plane. We both boarded the plane, but Wichman never spoke to me. He later told me I was politically unacceptable. I wrote him a letter informing him that his decision was an outrage, and reminded him that he had accepted a notorious Republican to head the downtown redevelopment drive without permitting the Democratic or Charterite minority to exercise a veto. Wichman subsequently called and announced that nobody ever talked to him that way. I told him, "I'm the guy." He asked me to come to his office to consider accepting the position. "Wichman," I said, "you come to my office. I'm not walking to City Hall again!" He came, and I took the job.

I received very little in the way of a charge to the Task Force, and even less with respect to my role as chairman. I was told, simply, that the Task Force was to be a planning body whose major task was to find out what type of community those people in the West End wanted. But I assumed from the beginning that I faced two basic problems. First, I expected that the blacks would have no confidence in me, and I could not blame them for that. Second, I suspected that the whites on the Task Force, and especially the city officials, viewed me as a "Patsy," someone they could twist around to do whatever they damn well pleased. The second problem bothered me least. I made winning and retaining the confidence of the blacks my top priority.

A chance to begin establishing my credibility with the blacks occurred very early, for I almost immediately came across an opportunity to demonstrate that I could get things done. At one of my first meetings with the Task Force the residents began to educate me, and the city officials, about what was wrong with the West End. One woman complained that a number of children and adults had been killed in traffic accidents at a certain corner. She added that they had protested and written letters requesting the installation of a stop sign, but nothing happened. After the meeting I went to Wichman, who referred me to the appropriate office. The man said there had been accidents at the corner, but no deaths. I said put up a sign anyway. He answered that it would cost $25 to $30. I told him I would pay for it. He said he would talk to Wichman. So I called Wichman and told him I would pay for the sign. He said no, the city would put it up, and in three days the sign was up. That started the confidence.

Next, it seemed important to me that the Task Force as a group score an early success. So I pushed hard to make sure our first project worked out. We decided to clean up and help restore Dayton Street, a street lined with beautiful old row houses dating back to the time in the nineteenth century when it was one of the city's most fashionable residential districts. The city possessed an expensive plan for it which would cost $200,000 or so, but lacked the money. So we got $10,000 from the Federation of Garden Clubs, or City Beautiful, or one of those outfits, and I extracted $5,000 from the city. The residents did the rest, or most of the rest. Somebody designed an arch, and Bill Baude, a representative of one of the city departments on the Task Force, picked up stone for it from a building torn down in another part of town. We persuaded every resident of the street but one to paint their houses, and the kids at Bloom Junior High built flower boxes which the residents picked up, painted, and hung under their windows. When we finished we had a big dedication, complete with the Bloom band, speeches, and all the rest. By this time the Task Force had momentum and we decided to tackle Queensgate II next.

All along, as part of the effort to do something—to create an aura of success—I had been working hard to develop a close rapport with city officials and city councilmen, a process which continued as we took on Queensgate II. That proved relatively easy. I just established regular lines of

communication and made friends with those guys. I used the usual techniques, including lunches at which we discussed the progress of our work
and what we needed from this or that official or agency. I spent endless hours
on the phone, especially with Peter Kory of the Department of Urban Development, urging people to be patient, explaining that the Task Force was
green and had to educate itself about the problems in the area and about
renewal politics, and assuring everyone that we had a good thing going and
we were going to pull it off.

I also checked in with businessmen, like Mark Upson, who chaired the
group which steered through downtown renewal, Dennis Durden, the city
affairs man at Federated Department Store, and Joe Hall of Krogers, just to
see how I was doing. They helped when they could, but were not very much
interested. After all, we were working the corner of Linn and Cutter, not the
corner of Fourth and Vine. They did get a little nervous when we talked
about going over to the center of Elm Street, because that put us in the central
business district, and Federated, with a big store (Shillito's) near the West
End, feared they might end up with a big, poor, black community right on
their front doorstep. Federated didn't want to be identified as the department
store of the poor community.

We also "prepared" people for our moves. Hugh Guest, a black, proved
invaluable at this. As the Queensgate II planning process began to take
shape, Herb Stevens of the City Planning Commission recognized that he
and his people had no experience with working with the community in a
planning situation. Hugh was in a seminar at the University of Cincinnati at
the time, and one night he told Mim Geiser about his work on a project rather
like ours in San Diego. She told Herb, and he took on Hugh and assigned him
as staff to work with me and the Task Force. Hugh agreed with me on
something I learned from Mark Upson. When decisions had to be made, we
decided to be sure they were made before the official meeting. So when we
had something to do I would ask Hugh if he thought we could put it over. He
would say yes or no, and then he would go around and check it out with two
or three key people, and I would do the same. We lined up our ducks before
the vote. Without that kind of preparation I would have been knocked on my
rear end when I walked into a Task Force meeting and asked for $300,000 to
hire Tom Jenkins and the University of Cincinnati. But I also "prepared" the
Task Force for Tom and the university by "accidentally" dropping a picture
of Tom during one of our meetings. The residents had not been especially
enthusiastic about Tom or the university until they saw he was black. That
discovery made the decision much easier.

Our early successes and our careful political preparation helped the Task Force, within a year, to develop unbelievable, almost frightening power. We functioned almost as a little City Hall. Nothing was done in the West End without our OK, and we almost always got what we asked. One of the sources of our strength was that the city learned we were serious, reasonable people trying to do a good job. And it meant a great deal to city officials. They did not have to do the work themselves. Besides, they were not equipped to do this kind of work, and they knew it. If there had been different kinds of people on the Task Force, though, the power could have been badly misused.

The Task Force worked well as a group. We rarely had any knock-down, drag-out fights along racial lines. Once, however, a white city official got up and made an ass of himself by saying some of the West End residents had not behaved as responsible citizens during the riots. He was referring to a few broken windows and a box factory that was torched. Rick Lewis took him apart. Lewis pointed out that just before the riots the factory advertised for a janitor. Right across the street lived an unemployed black man with seven kids. He went over to see about the job, and the factory refused even to interview him. Rick said the Task Force should think about that man watching 300 people go to work every day in that factory, suggested that maybe that explained the burning of the factory, and challenged us to wonder what kind of good citizens ran the factory. That was the most serious breech we had, and it blew over. Most of the other arguments occurred over planning, and while I let people talk—people have to shoot their mouths off—I also rode herd. Sometimes I cut them off and refused to let them speak all day. You have to do that if you're chairman.

Most of the serious divisions on the Task Force broke not along racial lines, but pitted the private sector against the public sector. The residents simply distrusted most agents of government, and often for good reason. I made it a practice that anything affecting the West End had to be discussed before the Task Force, and that set up one of our liveliest public-private disputes. One day the director of OKI, the regional planning agency for the Ohio, Kentucky, Indiana metropolitan area, called Peter Kory and me and asked us to lunch to discuss OKI's plan for a cross-basin expressway which would run through the West End. I told him the residents would not like the idea, but invited him to appear before the group. They tore him up! They were not rude, they just did not want it. He never understood that; never understood, apparently, how the history of the West End's experience with I-75, Lincoln Parkway, and busy streets without stop signs made road construction such an emotional issue. Like certain other planners, he thought he could do whatever he wanted.

Richard Krabach, Wichman's successor as City Manager, also posed a problem. It seemed clear to the residents that he did not like the process we had set up or the federal government's Neighborhood Development Program. I had to cultivate him and he was a difficult man to cultivate. I finally got to him, but it took a lot of time and hard work. I told him his attitude was very much anti-citizen participation and machine-like. All the while I urged him to come down and talk to the Task Force, but he was always busy. He finally agreed, though he expected to get murdered. I told him to be intelligent, and reminded him that he was City Manager and we needed him. He gave a good, logical speech in a candid, pleasant way. The Task Force asked questions, but the residents did not tear him apart.

One recurring private-public sector conflict stemmed from the residents' deep and persistent suspicion of the Department of Urban Development and its director, Peter Kory. Some of that grew out of the city's traditionally slow response to West End problems. Some of it came from the way urban renewal wrecked the West End. The residents tended to hold Kory's department responsible for all that. So whenever Kory, as a professional, stood up and said what ought to be done, or recommended a consultant, the residents said hold it, and looked at the proposal very carefully. Usually, both the public and private sector sides on the Task Force functioned in these collisions through caucuses. Before a vote on a specific issue, the residents and the city people would caucus to make sure their forces were together. Hugh tells me Mallory was probably the force behind the residents' caucus, though he seldom participated in Task Force deliberations. Rick Lewis and Jerome Jenkins ran *that* show. And they always made sure all their people were present for a close vote, because they could control 50 percent of the vote on the Task Force and could block or delay a question until they fully understood and agreed to accept or reject it.

The caucus and voting system, combined with my policy of hearing people out, took time. Occasionally we spent a whole day caucusing and listening to criticism of the city. Kory often got very frustrated. He would call me up and complain about wasted time or a "bad" decision. I finally said to him, "That's your job! You're getting paid for it. It's part of the game!" He learned, too. In fact, all of us learned a helluva lot. But Kory was very impatient because he was very busy. Stevens and others seemed to grasp the process sooner. In any case, I got very little pressure to try and control the criticism of the city.

Hugh Guest also helped overcome the private-public sector split. He spent hours working with Doris Brown, who turned out to be one of the most influential of the black residents on the Task Force. She really brought the West End community along. The first time we called a meeting of the West End community, about 2 people showed up. We asked Doris to help, and we had 124 at the next meeting. She simply walked down the street and told everybody she saw, "You're going to a meeting." All of a sudden we had 124 people. And they stuck with us. After the university was on board, she and others in the community realized they were involved in a significant way. They came to meetings once a month or so, usually held in the basement of a church, to listen to the planners and the experts. The residents sat down and said, in effect, "you planners are going to explain what is going on, and you are going to keep on explaining until we are ready to go to bed."

But Hugh's work with Doris paved the way for all this. Every Friday from three to five in the afternoon for a period of three or four months Hugh sat down and talked urban renewal and planning with Doris. He went over the city's 1948 master plan with her, and tried to give her a larger picture of where our project fit into the scheme of things. He also explained problems he saw with particular proposals for Queensgate II. Once she began to understand things, she felt less threatened by urban renewal and experts, and she would readily challenge people. So Doris, like other blacks on the Task Force, helped educate the community. It was beautiful.

In fact, our commitment to involve a broad variety of people made the Task Force go. We had an "augmentor" mechanism by which city council-men and other city officials could sit with the group. For councilmen, we chose people from the Urban Development Committee, so they knew the pros and cons on issues well before proposals got to committee, let alone Council. Some nonelected city officials were also augmentors, and when-ever they or other augmentors said they could not come because they were too busy I got on the phone. I said, "Well, you're no busier than I am, you bastards; you got me into this job, now come down and do it." And they came. But the point is we had people on the Task Force who were representa-tives of the residents, city departments, Council, business men, the Catholic Church, and the Protestant churches. We also had the biggest landowner in the West End. We had, in short, all kinds of interests and groups repre-sented.

The range of contacts of Task Force members helped immensely on a variety of occasions. One of the dramatic examples of that was the way we tied Music Hall renovation to the Queensgate II plan for a Town Center adjacent to Music Hall. I happened to be board chairman of the Cincinnati Symphony Orchestra, and Music Hall, which is owned by the city and houses the orchestra, was falling down in a dump. We wanted to rehabilitate the place. So we got $2 million from the Corbett Foundation, a local operation, and went to Council for four installments of $500,000 each. Council did not put a very high priority on this, and we had to fight for it. I argued with council members privately, in committee sessions, and on the floor of council. I told them, among other things, that the city spent two to two and a half million dollars of taxpayers' money on Riverfront Stadium to keep professional baseball and football here, and we could spend a little for other things for the city too. We did fine for a while, but the last time we approached Council the members wavered. So we arranged a dinner before a concert. I sent handwritten, hand-delivered invitations to all council members and their spouses. Only one said he could not come. So I called him and said, "Look, you S.O.B., you're coming, and so's your wife." They came. To top the evening we had Councilman Myron Bush, a black and one of our strongest supporters to this point, read the Lincoln piece in the Benjamin Britten thing over the orchestra. We even gave him a dressing room with a star. He had a beautiful voice and performed great. And he was tickled to death. We got the money. Bribery? There is no law against flattery, and not a single penny changed hands.

With rehabilitation of Music Hall under way, Kory began to push the Town Center idea hard. He got the OK from Krabach, who was City Manager then, and it began to look as if we would have a Town Center with shops, art, and music studios, the kind of stuff needed to attract racially mixed housing around it. But no money came along and the Town Center project stalled.

Then we got a break, and again our wide contacts helped. After rehabilitating Music Hall, maintenance costs—air conditioning, escalators, and the rest—mushroomed. So Corbett offered to build a garage next door and give it to the city if the money from the garage went to Music Hall. There were some technical, legal problems with this, but a woman lawyer in Urban Development came up with an ingenious solution for those. At the same time WCET, the local educational television station, wanted to move. So WCET—I'm going to be on the board—secured a grant of $2 million from the Crosley Foundation and $500,000 from the Schmidlap Foundation, both local agencies, to put its studio and offices on top of the garage. That revived the Town Center. The school board agreed to establish a vocational school there, and we began to look for other tenants, including the University of

Cincinnati. There was to be a bridge from the garage to Music Hall, which is now one of the greatest concert halls in the world. I'm sure we will be able to get one or two big insurance companies, and possibly others, to start housing around there. We will have a helluva Town Center.

The wide range of contacts represented on the Task Force also helped us secure top-notch consultants. The University of Cincinnati was our principle consultant, even though it was a compromise choice. We had severe problems making that selection. West End residents on the Task Force wanted the City Planning Commission to do it. But City Planning could not do it because the City Manager would not give them the money. He would not give them the money because of the rivalry between the City Planning Commission and the Department of Urban Development, and, I suspect, because the Manager and the head of the Planning Commission hated each other. In any event, I could not get the Manager to let Planning do it. And Kory argued that private consultants, such as the firm that did the planning for downtown renewal, would not do it because the job was too big, and would occupy six or seven of their people. So I said to Kory, "Let's see if we can get the university involved, it ought to be involved."

Kory and I then went to Wichman, who thought using the university was a good idea. We arranged a lunch with President Walter Langsam, a couple of vice presidents, and three or four other guys. I talked about the project, and I could see that they wanted it. So did Wichman, but the residents still favored the Planning Commission. At one meeting we even discussed bringing in the university without Task Force permission. By this time I knew the university was about to hire Jenkins to do this job, and I believed his appointment, because he was black, might bring the residents around. But Kory and I figured we could not safely tell the Task Force and the public about Jenkins because the university might change its mind, embarrass Tom, and maybe get him fired from his old position. That was when I "accidentally" dropped Jenkins's picture at a Task Force meeting. It was unofficial, but it cleared the way for bringing in the university.

I think the arrangement with the university worked out well. The planning team did a good job for us, and I would recommend doing it again. That is not to say that we did not have problems. The members of the university team—except Jenkins, who had been through this before—were green. They had no more experience with this kind of thing than the rest of us. As a result they lacked confidence, they were slow, and they made mistakes. One of the first problems was language. They began by making presentations to this unsophisticated Task Force and to the West End community in the jargon planners use. None of us could tell what the hell they were talking about! It was very vague language. But they corrected that, and learned to say it in English so all of us could understand.

Kory and the university team also clashed occasionally. Hugh and I had the feeling that Kory felt professionally superior to the university people, and at one point he said he was going to recommend that we get rid of the university team. He wanted to do that because the university people seemed inexperienced, because he disliked some of the ideas coming out of the team, and, I suspect, because the team, unlike professional private consultants, had no track record of devising a plan that would be practical in the bricks and mortar sense. I reminded him that Council had signed a contract with the university, and asked him how he expected to get the $300,000 back. I repeatedly urged him to take it easy, we would get this thing done.

The split between Kory and the university eventually blew over, but in a curious way. One night about nine or ten o'clock, Kory, Guest, Tom Jenkins, and Hayden May met to go over some of the team's housing alternatives. As Hugh tells it, Kory preferred high-density, high-rise housing because it made economic sense, and presented very convincing arguments to support his position. The team preferred lower-density housing. They argued that it might cost more in dollars, but felt that we could compensate for that loss in social and psychological benefits. Kory kept raising questions and objections. Finally, May lost his temper. He jumped up, whacked the table with his fist, and shouted at Kory, "Damn it, Pete, if you think we can't do it, then tell us and get us off the job!" That seemed to impress Pete. After that he began to really deal with the university team, and to treat May and the rest as equals.

There were times, too, when university team ideas collided with the views of the residents on the Task Force. One of the first of these disagreements also concerned housing. We discovered very early that the residents wanted single-family housing with conformed to the TV stereotype of "middle-income" homes. Everybody wanted to make Queensgate II a suburban neighborhood, complete with houses surrounded by yards and fences. It took a long time for us to explain that suburban development was economically impossible in Queensgate II. A trip to Washington helped on that. At one point HUD held up on the reservation of funds for our project, so Mallory, Kory, and I arranged to take some residents to Washington to protest to appropriate HUD officials. We succeeded, although somebody told the mayor we took a bunch of communists to HUD and created an awful impression. I told him, "Yeah, they were terrible people, but we got our money." In any case, while in Washington, we took our group to Reston and some of the integrated developments in the area to show them the kinds of things we might do besides build a TV suburbia. The residents finally went along with the university team on that issue.

We had a similar split over education. The university team talked very early about getting vocational education into the plan. That upset the residents. The problem was that people in that community thought of vocational education as training to be a barber, a hairdresser, a cook, or a "plumber" who repairs 100-year-old plumbing in crummy tenements for low wages. Hell, plumbers and carpenters who did big jobs in subdivisions then made $20,000 to $40,000 a year and drove around in Buicks and Cadillacs. But inner-city people never saw *that* kind of plumber; for them, carpentry, plumbing, and factory work seemed demeaning occupations.

On the other hand, the university team misunderstood many of the residents on the education question. Residents liked to talk about the need for quality education, not only for their own kids, but to serve as a magnet to attract middle-class families into the neighborhood. When some university team members heard this "best school in the city" talk, they assumed it meant a high-powered academic school which would, they felt, isolate the middle-class kids from the poor kids. Both sides agreed we had to put in "good education" to secure a mixed population, but there was lots of misunderstanding on both sides about what "good education" meant. I hoped the Town Center, with WCET and the Board of Education's vocational facility, would enable us to develop a flexible program with high-quality academic and vocational content.

Other consultants played a lesser role, and we had little trouble with them. We managed to get good people because our selection process worked so well. A committee, which I chaired, interviewed and screened potential consultants and recommended one to the Task Force. Kory was especially valuable to us. He would present us with a list of possibilities based upon his really vast and impressive knowledge of this new industry. He always picked good ones, and we had only one flare-up. Kory sent us a list of three or four firms to handle the job of working out the details of the Town Center. Kory clearly wanted us to choose a firm he had worked with on another job in Cincinnati. But most of the committee liked Design Associates, an outfit from Pittsburgh. It sent one of its top people to the interview; he was warm and sympathetic to blacks and underprivileged people. And he gave us no baloney. But, more importantly, and even though he had moved from England to America, he communicated beautifully with West End residents like Doris Brown. You could not find two more different dialects of the English language.

We almost had a fight, literally, before we settled on Design Associates. Neither Hugh nor Kory served on the selection committee, but both attended our decisive meeting on this particular consultant. Hugh and Kory got into a hot argument. Kory was coming down strong for the firm he worked with here before, and Hugh was pulling for Design Associates. At one point, Hugh complained that the firm Kory wanted did not have the good sense to include an established neighborhood bar in its plan for the other Cincinnati neighborhood. Kory responded by asking if the city should develop a policy which recognized that bar. Now, Kory loved to ski, so Hugh answered by saying that if the city could have a Director of Urban Development who stood on top of the Alps with 1 × 4s on the bottom of his feet the city could have a policy which recognized that bar. The argument continued, and finally Hugh threatened to come around the table and hit Pete. I pulled him back and suggested that both of them leave the room so the committee could make its decision. We chose Design Associates, and Kory got mad as hell at me. But he later agreed we made a good choice. I think he is now very happy with that decision. Hugh soon got over his anger with Pete, and they worked OK together. Ironically, about a month later, Pete fell while skiing and broke a leg.

We had more trouble with developers, though we selected them in the same way. We had three or four developers, including a couple of local people, who wanted to take on the biggest part of the work. Kory got Mid-Cities, a nationally known firm, to enter the picture because he knew the head of the firm when this guy worked for HUD. So the committee sat down to listen to the sales pitches. Mid-Cities sounded by far the best, and we went for them. Then we had to go before the Council Committee on Urban Development, because two local outfits which lost out requested a special hearing. One guy accused me of taking a bribe and said the committee had been subjected to undue influence. You always catch hell from losers. Council eventually went along with us.

Finding a developer for Block "C" proved especially painful. We decided to pick Ben Schottenstein, who had owned a department store in the Queensgate II area for years, over two other people. Ben had an engineer do his plans, and they were not very good. But we picked him because he offered to subsidize small businessmen by charging them $3 rather than $6 per square foot. We also stipulated that he should get an architect to rework his plans from a list of 6 or 7 provided by the city. As it turned out, he picked one who lived outside the city. That created a problem. One day when I was driving to work, a councilman pulled up beside me at a stoplight. "God damn it," he hollered, "you could at least pick local architects. You make it tough on elected officials." So I said, "Was that silver Porsche you're driving made locally? My cheap Chevrolet was!" That shut him up. He laughed and dropped it. We had to get an outside architect because, in the opinion of the city of Cincinnati, there were no nationally known architects in town who could handle a job like this.

But that did not end our problems with Ben. The two losers claimed we changed the rules in the middle of the game by letting Ben change architects, and they began to get to other local politicians. One night Corbett had a dinner for all the council members, and Ted Berry and Taft pulled me aside. They said, "Buddy, we're afraid we will have to turn you down on the Schottenstein thing and open it up again. We hope you're not personally affronted." I told them I was not affronted. But I added that if they reopened it we would never get this done because more people would enter and it would take two or three years to decide. I argued that a delay would not be fair to the people in the West End. They did not care who did it, they just wanted it developed. The council members did not seem impressed. So the next day I called up each one. "Look," I said, "we didn't change the rules. Everybody had the same right to say he would subsidize the black merchants. Schottenstein said he'd do it; the other guys didn't. That's why we picked Schottenstein." To add a little frosting to the cake I also told them Schottenstein had plans to attract AVCO Broadcasting, which wanted to move, and Greyhound, which, I said, would move to Newport or Covington if we did not act soon.

Still, I got no definitive assurances. Instead, the Urban Development Committee referred it to Council. So I went to the Council meeting, where I initiated discussion of the issue by talking a lot. Finally, Berry led off the debate. I didn't know where he stood, although I believed all the Republicans but one would come along. Berry said he was seriously considering turning Schottenstein down, but in reviewing the case and so forth . . . He talked for about half an hour, and concluded that he now favored Ben. Taft agreed, and that did it.

Ben was delighted. Afterward, in the hall outside Council chambers, he kissed me on the cheek. "You need three suits?" he asked. "Ben," I replied, "one of my suits costs more than your whole inventory!" He laughed and thanked me. A little later he called me up. "No kidding," he said. "I want to send you something." "Ben," I answered, "the minute I ring off, after

thanking you, I'm calling Bob Turner, the City Manager, and telling him. So
don't do this. He'll call the whole deal off." Ben never sent me a thing. I
could not have sold it if he had.

We had one other problem with a potential developer. There was an
ecumenical organization in the West End that wanted some land for its own
purposes. This group was interested in rehabilitation, and it had an architect
and a lawyer. There was a group of black militants around—the Black Turks,
or something like that—which used the facilities at the DePorres Center.
One day they came down to talk with Hugh about getting some land. Then
they invited him to come and see them. They had a meeting on a dark, rainy
night in an old garage. So Hugh walked in and explained that he worked in
planning, that he did not buy land, or demolish buildings, or any of that.
They dropped it after that. The group might have been happy to cooperate,
but the Queensgate II Club, WEDCO, and the Task Force never invited
them in.

So we finally got the plan together and the developers lined up. But we
accomplished more than that. The Task Force and the community participation process also produced sophisticated leaders. First Bill Mallory and then
Jim Rankin moved into the state General Assembly from the base and
reputations they built in the West End. Hugh Guest went from our staff to
become director of CDA. Jerome Jenkins moved on to be head of Seven
Hills Neighborhood Houses. John Jackson got his start in community work
with the Task Force, spun off from that, and did a real job in getting a
building up for the West End Health Center. It was a great educational and
growing process. We had a 70-year-old black woman with us in Washington
who stood in front of a HUD undersecretary's desk and told him, "Dammit,
you better get some housing built down there before I die." She would never
have done that before she had experience with the Task Force and the
Queensgate II Club.

Hugh is convinced that this experience molded the West End into a strong
community and made the Task Force into a strong and distinctive community organization. People in the West End are poor, but they are now comfortable in dealing with the business types and others they got to know
through the process. They have power, and they do not hesitate to tell the guy
who wants to put up an office building, "OK, this is what you have to give up
if you want your building down here." But Hugh is absolutely sold on not
seeking a program to help poor people through an organization dominated by
poor people. As he puts it, "You don't want a poor people's organization,
because you'll end up with a poor people's program." You need a variety of

influential people participating with poor people. And before you know it Doris Brown is influential, too, because of her participation and contacts. You do not want a poor people's program. You want a people's program.

I think we also produced something for the university. The university has one of the biggest budgets in this town, and it ought to be intimately involved with the city, not off in a so-called community of its own. We were comfortable with their work and the plan, which we helped formulate by discussing and voting on all the alternatives, block by block. It was good for the faculty and the university, even though it created a helluva turmoil up there. People still want to talk about it. Yet, afterward, the guys on campus tended to retreat. They did some excellent studies, but the university administration did not seem to take them too seriously, so we downtown began not to take them seriously. The way to change that is to get them off the campus, mixing with all sorts of people. And people with experience in the community ought to be up there, mixing in the classes and seminars. Everybody is too isolated. I think it is more important for the university to be involved with the city than for the city to be involved with the university, but the two have to work together.

PART III

Research and Planning:

PERSPECTIVES FROM THE UNIVERSITY TEAM

The essays in this section assess the role of the university's research and planning team in the Queensgate II urban renewal project. In the first, Hayden May, who served on the core planning staff, and Zane L. Miller, who served on the social science research group, analyze the outcome of the team and Task Force deliberations on the plan's housing component. In this view, the process failed because the housing policy guidelines hammered out by the Task Force contained "loopholes" subversive to the Task Force's previously adopted housing goals. The culprit, May and Miller contend, was to be found in the circumstances which stretched the process of guideline formulation to the point that the exhausted and frustrated participants accepted what seemed expedient instead of pursuing the guidelines argument to its best possible conclusion.

Harry Dillingham, the author of the second essay and a sociologist who served on the social science research group, also finds the plan defective, but from a different stance. He criticizes the entire process as intellectually unsatisfactory and contends that both the housing and educational components of the plan rested on inadequate research and therefore the shakiest of social science foundations. From his point of view, the plan is unrealistic and, because of the nature of the process, could not have been otherwise. That contention stems from Dillingham's belief that irresistible social forces shaping the metropolis ran counter to the basic objectives and assumptions inherited by the university team and from which the plan was built.

The third essay is by Lewis Bayles, the educator on the social science research group, who deals with the difficulties he encountered in his area of special responsibility. Unlike Dillingham, he faults neither the process nor the plan, concluding instead that only time will tell. In a fundamental sense, and unlike Dillingham, Bayles conceives it possible that what people do is more important than the influence of social forces, and his essay centers on how it is that people did what they did during the Queensgate II planning process.

In the final essay, Thomas H. Jenkins and Jayanta Chatterjee, director and assistant director, respectively, of the university team, review the tribulations they encountered in managing the university team. They tell us, too, that they aimed not merely to complete a plan but also to produce the kind of plan that would help turn urban renewal in new directions. In a self-conscious attempt at writing a "balanced" assessment, they concede that they won a few and lost a few, but argue in the end that the project achieved both their aims and that it was, therefore, successful.

CHAPTER 6

Housing:

THE CRITICAL NEXUS

Zane L. Miller
Hayden B. May

Housing proved the most difficult and the most volatile question in developing the Queensgate II plan. The issue had to be faced within the context of guidelines laid down before the university signed the planning contract. Central business district interests and individuals in the city's planning bureaucracy harbored predetermined conceptions of the appropriate housing program which either implicitly or explicitly contradicted those guidelines, and both the team's planning staff and the social science research group repeatedly split in the process of working out goals, programs, and policy recommendations for consideration by the West End Task Force. More frustrating still, the final policy guidelines on housing adopted by the Task Force and City Council contained loopholes which might, during the implementation stages of the project, wipe out the essential features of the housing plan worked out by the university team and adopted by the Task Force.

Initial planning statements for Queensgate II underscored the overarching significance of housing in redeveloping the area, but did not establish either specific or irrevocable sets of objectives. Between 1965 and 1968, before the university planning team began its work, the West End Task Force, the Model Cities Agency, and the City Planning Commission and Urban Development Department of Cincinnati went on record in favor of making that part of the West End which lay east of I-75 primarily a residential neighborhood for an economically diverse and black, predominantly black, or integrated population. Federal policy guidelines suggested, but did not absolutely require, that renewal projects adhere to these goals to qualify for federal funding. A Department of Housing and Urban Development directive dated February 1968 stipulated that, wherever feasible, renewal projects should reduce the excessive concentration of minority-group families and foster equality of housing opportunity throughout the urban community. The same directive, however, took account of "overriding considerations" which might prevent the dispersal of minority concentrations or the broadening of housing opportunities, and permitted the undertaking of such projects if the city had "other plans" for achieving the same ends. This qualified endorsement of fully integrated neighborhoods opened the way for contention among the several interests and individuals seeking to influence the direction of the Queensgate II housing plan.

None of the parties to the contest sought very seriously to overturn the commitment to residential as opposed to commercial or industrial development, but it became clear very early in the planning process that several important groups and individuals hoped to push the housing plan in different directions. The planners on the university team soon recognized that they faced a conflict situation and decided to try to coax the contending forces into compromises. To walk that rope the planners developed sets of alternatives which reflected the various positions and then referred those alternatives formally and informally to the involved parties in an attempt to gradually narrow the range of differences.

Developing and securing adoption of a set of housing goals loomed as the initial task confronting the university team. But before approaching the Task Force with a goals proposal, the core planning staff and the interdisciplinary faculty research team spent the summer of 1968 researching the neighborhood, interviewing interested parties, and exploring alternatives. By September these preliminaries had been completed, and during that month the staff developed two preliminary goals statements. The first of these, formulated on September 19 and designed to provide an initial working draft for consideration by the planners, drew on previous guidelines prepared by the

Task Force, the Model Cities program, and reports prepared by the core staff and the social scientists. Though generally phrased and bearing a decidedly "motherhood" ring, the document opted for developing a mixed residential neighborhood. The statement fell into three parts. The first suggested the elimination of all housing which had deteriorated beyond the point of architecturally or historically justified rehabilitation. The second plumped for the construction of new housing designed to attract an economically and socially heterogeneous population. The third urged the adoption of provisions for minimizing the adverse effects of relocation on current Queensgate II residents.

Within a week, however, and before going to the social scientists for consultation, the core staff had reformulated the September 19 goals statement. The new proposal eliminated the "motherhood" phraseology about the creation of an integrated neighborhood. Instead it cited "constraints" that might prevent the establishment of that kind of neighborhood. Given the prevalence of poverty in Queensgate II, its location adjacent to a poor residential district, its functional isolation from the larger communtiy, and its sparse population, the team concluded that the September 19 goals appeared decidedly visionary. But the report did not end there. It added that these constraints might be overcome if income levels and population in the area could be increased, and if new functional ties to the larger community beyond could be forged. The core staff statement of September 26, in short, foresaw the improvement of the Queensgate II environment to make it attractive to "new users," the establishment of activities in and adjacent to Queensgate II which supported that end and which also served other elements in the metropolitan areas, and the strengthening of relationships between Queensgate II and the larger community to open social, economic, educational, and recreational opportunities for local residents beyond the bounds of the immediate neighborhood. It concluded by suggesting that these three goals be given top priority in shaping the Queensgate II plan.

After discussing this paper with the social scientists, the planning staff, early in October, prepared a revised goals statement and submitted it to the social scientists for criticism as a preliminary to presenting it to the Task Force. The draft, eighteen pages in length, contained an elaborate explanatory apparatus, including three appendices, one of which contained definitions of terms commonly used by professional planners but which struck the social scientists as too jargonistic for public consumption. The draft also more fully presented the case for attracting new users to the area. Since

Queensgate II began to die as a residential neighborhood after it had been cut off from the rest of the community, the argument ran, the area had to be opened up either to West End or metropolitan-scale use. With respect to housing, this meant that unsound dwellings should be eliminated and new units should be designed to meet the needs of households of various sizes and income levels while retaining top priority for current residents in the area who wished to stay in securing access to the new facilities.

At this point simmering disagreements among the planners and the social scientists about the housing goals burst into the open. Some of the planners expected that this early October statement would, with minor revisions, go on to the Task Force for its consideration. But the social scientists objected to sending a goals statement to the Task Force without benefit of more intensive faculty review. Since, moreover, the social scientists divided on the question of housing, particularly with respect to the kind of population mix that might be induced to remain in or move into Queensgate II, regardless of the kind of housing in the area, the planners determined to devise yet another goals statement. At this stage the planners adopted a strategy of dealing with the divisions among the social scientists by writing up alternatives which reflected basic divisions over housing policies. By the end of the month the first report utilizing this approach had been prepared and stood ready for discussion with the social scientists.

The October 30 goals statement encompassed three alternative directions covering a spectrum of development possibilities ranging from exclusively residential to nonresidential land use patterns. The planners deliberately sought to make it as comprehensive as possible and included within it statements which seemed palpably infeasible. The objective was to open an extended discussion with the social scientists so that all members of the university team would feel satisfied that they had seriously considered all possible solutions to the problem of Queensgate II's future. This document, more than any of the earlier statements, revealed the complex of forces converging on the planners and indicated the various visions of what the area should become.

The first alternative consisted of 16 pages, focused on residential development, and contained 7 possible income and racial mixes, each of them worked out at low-, medium-, and high-density levels. The second considered various ways by which 30 to 50 percent of the area could be devoted to West End-related (that is, black) residential development and the remainder to region-serving activities, including those related to the central business

district, the West End, or the entire metropolitan district. The third provided
for the housing of all current Queensgate II residents outside the area and the
installation of commercial, industrial, recreational, health, welfare, civic,
educational, or some mix of these and other institutions which might serve
either the central business district, the West End-Basin area, the entire
metropolis, or some combination of these regions.

The university team discussion on the night of October 30 reduced the
range of alternative directions by eliminating the extremes of exclusively
residential or nonresidential land use patterns from consideration. The following day the staff put together a compressed statement of the remaining
alternative goals and in the following weeks circulated the paper among all
interested parties, in part to bring people beyond the university up to date,
and in part to test the direction of nonacademic opinion.

The response suggested that basic positions had not altered significantly.
Representatives of Queensgate II residents still thought primarily in terms of
restoring the neighborhood for their use and housing its population in single-
family detached homes set on separate lots. Central business district interests, especially department store representatives, preferred replacing low-
income families with high, a position which looked toward high-rise or other
luxury apartment construction. The planning and development bureaucracy
in City Hall seemed inclined toward restricting housing redevelopment to a
minimum and devoting as much of the area as possible to institutional uses
serving the central business district, the West End-Basin area, and the metropolis. The university team remained divided and ready to fall back on the
strategy of presenting a wide range of alternative goals to the Task Force.

Throughout November the planning staff wrestled with the problem of
preparing a goals recommendation for consideration by the Task Force in
December. Encouraged by Peter Kory of the city's Urban Development
Department, the planners decided to confront the Task Force with several
alternatives in order to place that body in a position of having to make
decisions rather than merely ratifying or rejecting one proposal. By December 5, after eight months of research, discussion, and lobbying, the goals
statement for the Task Force was at last ready.

The alternative redevelopment goals for Queensgate II presented to the
Task Force on December 5 seemed astonishingly bold. None precisely fit the
desires of the groups outside the university circle of planners and social
scientists and the statement narrowed the range of housing alternatives
precipitously from the broad spectrum which the social scientists agreed

ought to be submitted to the Task Force at the October 30 faculty-planning team meeting. The three options centered on whether Queensgate II should be opened to closer ties with the West End and Over the Rhine, with the West End and the central business district, or with the West End, Over the Rhine, and the central business district. But in each instance the housing goals recommendation called for mixed-income residential development and reserved 50 percent of the area of residential use. Each of the three alternatives, however, was presented as a direction that required further study. The Task Force approved the report and agreed to vote on the planning goals at the meeting scheduled for the following week.

On December 12 the Task Force concurred with the university team's basic diagnosis that isolation had killed Queensgate II as a viable residential community, but backed away from the planning team's housing recommendations. The planners presented eight questions to the Task Force for a vote. Six of them, dealing with extending the functional relationships of Queensgate II to various combinations of adjacent areas and land use criteria, sailed through unanimously. The two questions on housing also secured a unanimous vote, but not until amended in a critical way. As presented to the Task Force, the housing goals requested authority for a plan which would include housing for low-, moderate-, middle-, and upper-income families and individuals. But the Task Force amended the critical imperative "shall" to the imprecise permissive "may." By these actions the Task Force endorsed the planning staff's argument that Queensgate II constituted a neighborhood dying of isolation but refused as yet to take a stand on what kind of people ought to occupy the new and more open community. The hard questions of Negro removal and social, economic, and racial integration remained to be faced by those beyond the university.

In January of 1969 the planning process lumbered from the goals stage into the development of a program for housing and related uses. The focus of action turned back to the planning staff and the social scientists, though an outside economic consultant, the school board, and the city's Urban Development Department played important parts. By the end of the month the planning staff had constructed three alternative directions for presentation to the faculty team. That document produced a fist-banging discussion and a fusillade of memoranda which all but destroyed the interdisciplinary complexion of the university effort. Housing and population characteristics constituted the critical issues in these exchanges.

Throughout most of January the housing specialists on the planning staff juggled figures dealing with dwellings unit densities and the age and income characteristics of the current Queensgate II population and projected new residents. By the end of the month three proposals had taken shape. Two of them allotted 40 acres for residential use and provided housing suited to the profile of the current Queensgate II residents. But one of these two assumed that 1968 market conditions in the area would only attract new residents who would demand one- or two-bedroom units suitable for childless couples, while the other, based on projected population of between 6410 and 7424, assumed that steps would be taken to develop an "economically balanced family residential area," and suggested vaguely that a significant expansion and "change" in educational facilities might be one of those steps necessary to attract families with children into the area.

The last alternative also reflected the growing conviction among the housing planners that a "strong" residential component could not be built into the overall development project unless additional acreage became available for residential use. It proposed working with 60 acres for housing, the additional 20 to be gathered from the Garfield Place and Washington Park areas, providing housing for the current Queensgate II population, and taking "whatever steps . . . necessary" to produce an economically balanced family residential area, including effecting substantial changes and additions to educational facilities.

On February 7 the core staff and social scientists met in a wide-ranging, frequently heated, and inconclusive discussion of these proposals. A week later the planners came back to the faculty with another formulation of the three proposals, now labeled X, Y, and Z, and spelled out in more detail the implications of each. By these calculations X would house 2716 people in 1280 dwelling units on 40 acres and provide for a population essentially like that already ocupying the area, but it failed to satisfy the goals adopted by the Task Force in December of 1968, calling for improved commercial services, upgraded educational opportunities, and income heterogeneity. Y would accommodate 7424 people in 2300 dwelling units on 60 acres, envisaged a mixed range of incomes among the population, and met both the commercial and educational goal criteria. Z foresaw placing 5726 people in 3000 dwelling units on 60 acres, fell short of meeting the educational goal, and rested on the attraction of 2-person adult households of undetermined income range and distribution.

By establishing these proposals, the core staff held to its strategy of developing positions which reflected the desires of the major participants in the planning process. In this perspective, X represented the wishes of the resident community, for the idea of single-family housing for the remaining inhabitants of Queensgate II still ran strong among their representatives. Z, on the other hand, appealed to the city, or at least to its most aggressive representative, and to the central business district interests. That alternative not only required few additional city services, but it clearly implied, or at least opened the door to, the construction of high-rise luxury apartments fitted out with aesthetic and recreational amenities to lure a large number of middle- and upper-income families with few children. It meant, in short, an increased tax base for the city and a minimal strain on the city budget for new services, and it constituted a contribution to the revival of downtown, one of the major aims of much of the post-World War II planning and redevelopment effort in Cincinnati. The core planning staff preferred Y because it dovetailed with their personal values, matched their concept of what a viable central-city neighborhood ought to be, and fit the analysis of what had gone wrong with Queensgate II that seemed most persuasive to the planning staff. Still, the planning staff did not take an advocacy position with respect to alternative Y. Working with all three alternatives, they felt, the Task Force could surely reach a compromise agreement with which all those most directly affected by the plan could live.

At this point, however, the university team plunged into another internal wrangle. Several social scientists still balked at forwarding to the Task Force any proposal, even when posed merely as an alternative, which they could not, as a matter of conscience and professional integrity, approve. In the next few weeks at least one member of the planning staff took the same stance. The intramural contest among members of the university team that followed constituted, in the final analysis, a challenge to the strategy of presenting the Task Force with proposals it could revise, combine, reject, or approve, and an attempt to break the university team's agreement, reached during a tumultuous February 15 meeting, to devote special attention to developing and supporting Y while at the same time working out X and Z as feasible fall-back positions.

The contestants fought the battle with memoranda. Between mid-February and mid-March, 4 hit the mail, 2 of them 48 hours after the February 15 meeting. The first, authored by a member of the planning staff and addressed to but 5 members of the university community, only 3 of whom held places on either the planning staff or the social science research team, bore the title "Proposal for Alternative X." The introductory rationale described it as a potential first stage in the gradual redevelopment and change of the West End, posited a 5-year implementation period, and took as its major goals the improvement and broadening of social and economic opportunities for present West End residents and "for new residents with similar socio-economic characteristics." The second, also dated February 17, came from a member of the faculty research team and expressed a marked distaste for the "gilded ghetto" potential in X. The third appeared 3 days later, went to the director of the project, and generally supported the "Proposal for Alternative X." Written by a social scientist, it warned of the dangers of placing the university team in the position of the characters in Clifford Odet's drama, *Waiting for Lefty,* and urged the planners to work out answers to the question of whether X could "go forward," either as a first stage in a bigger plan or as a complete plan, should a pending report of the economic consultant or other events lead in that direction.

The fourth memorandum appeared on March 12 and came from the same faculty member who had earlier associated X with the creation of a gilded ghetto. Addressed to "All the Q-IIers," it reminded them of the February 15 agreement to give scheme Y top priority, attributed the wave of race riots in the mid-1960s to the concentration of urban blacks in the inner city since at least the turn of the century, and contended that Y, with its high-quality educational program and heterogeneous population, raised the possibility of mitigating the polarization inherent in a metropolis residentially segregated by race. And if Y failed to attract an integrated population, it concluded, the result would be very much like X anyway.

Meanwhile, the director, assistant director, and core planning staff had been patiently explaining to their colleagues at the university that X, Y, and Z were not under review for a university team decision, but merely for the purpose of finding if they were "in the ball park," and at the same time conferring with the outside consultant on the economic feasibility of the three alternatives. By March 21 those discussions had terminated and the core staff had prepared a report on them for faculty consideration, as well as a fat packet of other documents exploring the age, income, recreation, and educational implications of each of the alternatives. All this information generally confirmed previous conclusions and tended to support the core staff's preference for Y. The economic consultant felt that Y could not work without providing "good schools" and "other amenities," a staff report made the case for using both superior schools and housing as bait to attract a heterogeneous population, and yet another staff report proposed a system of decentralized "satellite" schools scattered through the housing units for preschool through the first four primary grades to be integrated by teaching one grade only in each of the various centers, and the addition of a middle school and a secondary-level educational park to serve Queensgate II and other Basin communities. The Board of Education, moreover, seemed intrigued by this educational package, for it minimized new school construction out of local school funds by placing much of the cost of the new physical plant in the housing redevelopment budget and by adapting present buildings for use in the new complex.

Toward the end of April, the economic consultant submitted a housing market report. The firm which made the study knew something of the three proposals being prepared by the core staff, but based its analysis exclusively on market factors in the 1960s. The consultants' findings generally pointed in the same direction as the core staff, though the report pointed to substantial constraints limiting the market for middle- and upper-income dwellings in the area and recommended that lower- and moderate-income households be given primary focus within each of the three alternatives. The consultant also revised the distribution of households by income level. The firm allotted scheme X 39 percent low-income households, 23 percent moderate, 36 percent middle, and 8 percent high. Scheme Y got 22 percent low, 32 percent moderate, 36 percent middle, and 9 percent high. Scheme Z came out with 28 percent low, 25 percent moderate, 27 percent middle, and 18 percent high.

By this time the core staff had concluded that all three alternatives were "in the ball park" and felt it could recommend one of them as the most appropriate for Queensgate II. In April the planners prepared a set of drawings, a slide program, and three-dimensional models of each of the alternatives. Throughout April and May the planners met repeatedly with a variety of community groups and central business district and city representatives to pick up criticisms and to gather ideas for inclusion in the final presentation of alternatives and recommendations to the Task Force.

The persisting passion within the Queensgate II community for single-family detached housing continued to be a problem through this period. The team met this difficulty in two ways. In working out the three-dimensional models, the planners discovered that about 150 more dwelling units could be accommodated in the area than originally projected. This enabled them to develop a fourth alternative (XX) by replacing the 150 "extra" dwelling units that might have been devoted to town houses with single-family detached homes. At the same time, the staff arranged (in June) to take community leaders on tours of recently completed housing developments, including

Columbia, Reston, and the Williamsburg complex in Cincinnati, to show them the range of housing types represented in garden apartments and town houses. Most community leaders had never seen this type of home construction before, and almost all of them found much of it quite appealing. In any event, single-family detached housing disappeared as a serious possibility in June.

On May 22 the core staff presented the alternatives and its recommendation to the Task Force. The planners opted for a modified version of scheme Y. As it now stood that alternative provided for a total of 2000 families, including the 650 living in Queensgate II, and established an income mix of 40 percent low income, 27 percent moderate, and 33 percent middle and upper income, with a 50-50 split between families with children and childless households. Housing for the latter group, according to the recommendation, would be placed near the Town Center, a shopping, service, and recreation complex, and near City Hall and Garfield Place, while families with children would be distributed throughout the entire project in such a way as to "encourage community integration and meet housing market considerations." Other parts of the plan requested the educational facilities the team felt necessary to make the housing plan viable.

The Task Force went along with the core staff's recommendation, but instructed the team to break it down into geographic units within the project area and resubmit that for an area-by-area vote. The decision list, allowing an approve, reject, or defer vote on each area, came before the Task Force on June 5, though that document recommended a defer vote on the four satellite schools on the grounds that "educators" felt the idea should be explored further before making a decision.

On July 4 the Task Force voted on the 12 decision areas. All but one of the housing recommendations sailed through with overwhelming majorities. Two recommendations for the area bounded by Twelfth Street, Central Avenue, Court Street, and John Street received a defer vote. The first involved 154 low-density housing units. The second covered 129 medium-density housing units and elicited an advisory amendment calling for a feasibility study to lower the density by eliminating housing containing elevators for use by families with children. With these adjustments made, the university team prepared to terminate its connection with the Queensgate II project and the director and assistant director returned to academic life on a full-time basis.

Yet several factors combined to pull the university back into the picture.

Before the recommended plan could be implemented, it had to be put into more detailed form for adoption by the Task Force and City Council and for forwarding to the federal government for funding. An argument now developed over who should draw these "policy plans." Community spokesmen seemed to distrust the city planning bureaucracy on the grounds that the agency might make subtle yet substantial changes in the project while drawing up the policy plans. The city, for its part, insisted that by the terms of the contract the university team should do the policy plan. University officials, caught in the middle and unwilling to offend either the community or the city, decided that the team, the planning staff, or some of its members, should participate in this final step. By the end of the summer of 1969 Hayden May was back on the project on an hourly pay basis, charged with overseeing the making of the housing policy plan. By the time that task was finished, in the summer of 1970, serious damage had been done to the housing recommendation approved by the Task Force.

May finished a first draft of the housing program policy in December of 1969. It detailed acceptable housing types and laid down specific minimum standards and quality controls for their design, construction, and distribution, and indicated which should be wholly or partially subsidized. Providing for the geographic integration among various income levels of families with children proved the stiffest challenge. The guidelines set up two alternatives for each subarea, one labeled "feasible" on the basis of currently ascertainable housing market trends, and the other categorized as "desirable" and possible only if carefully synchronized with the development of superior educational facilities and staged so that early housing starts for families with children were not given over entirely to low-income dwellings. The former would have segregated poor families with children west of Central Avenue, and the policy guidelines devised by May and previous Task Force goal decisions and housing recommendations ran counter to this. In this context the December policy program virtually bound the city and the developer to the "desirable" alternative and the economic integration of families with children. This proposal went to Peter Kory for comment. He responded on January 15, 1970, approving the document generally, but objecting to several parts of it on the grounds that the lack of "flexibility" would make it impossible to find a developer.

He expressed concern that the guidelines on high-rise apartments for childless households went too far in limiting the height (10 to 15 stories) and number of units, but, more importantly, he maintained that all of the policy recommendations, and particularly those providing housing for mixed-income families with children and their geographic integration, should not be sent on to the Task Force and City Council until and unless the economic consultant certified that they had a "good chance of being achieved" and represented "a sound and feasible way" of fulfilling the pledge to heterogeneity built into the housing recommendations previously approved by the Task Force. May felt that Kory's suggestions for revisions, taken together, subverted the housing program by making it so flexible that the developer might change the entire character of the residential structure of the neighborhood.

On Kory's suggestion, May, Tom Jenkins, and Kory went to Washington to hash out the housing tangle with the economic consultant. May and Jenkins contended that they could not write a policy program which permitted the destruction of the entire plan, while Kory held out for increased flexibility to secure a developer and insisted that May must resolve the conflict between flexibility and control. After the discussion, May and a staff member from the consultant's firm sat down and drew up another set of guidelines. In the process they hit upon an ingenious though not infallible device to protect the opportunity to achieve both heterogeneity and income mix among families with children.

The revised policy guideline, dated January 15, 1970, aimed at keeping the door open for a developer willing to meet the desirable heterogeneity and distribution goal. The object was to avoid segregating low-income families from moderate- and middle-income families. It allocated low-, moderate-, and middle-income family housing for each subarea of the project in two ways, one meeting the economic consultant's estimate of marketability at the present moment and the other meeting the desired distribution to provide both heterogeneity and income mix within the Queensgate II project area. These figures stood as outside limits within which a solution could be found during implementation, but the planners had worked out an income mix schedule for each subarea by which the housing market could determine the character of from 25 to 50 percent of the units in each subarea and still produce income heterogeneity within the area. For this approach to work, however, a careful staging of implementation was required, so that early construction established the heterogeneity and income integration pattern for families with children. Persistent pressure from the community and the Task Force was also needed, to see that the developer and the city did everything possible to live up to the income integration goal and sought conscientiously to stick to the 25-50 percent market plan for each subarea. But the revised housing program policies did seem to resolve the conflict between Kory's insistence on flexibility to attract a developer and the planning team's desire for tight controls to assure the fulfillment of the heterogeneity and income distribution goals.

The policy recommendation for housing went to the Task Force on January 22, 1970. The planning team made no formal presentation at that time, but sought instead to convince the Task Force that the statement did not give the city and the developer the leeway to wreck the Task Force's housing goals and recommendations. By April a prospective developer appeared on the scene. Called the West End Development Corporation (WEDCO), it was a nonprofit organization made up of local components from the West End community, including some influential black political figures. For a time it appeared that WEDCO objections and community impatience would twist the Task Force plan into unrecognizable shape.

On April 9, H. Ralph Taylor, a former Assistant Director of HUD and a private housing and urban development consultant working out of Washington, appeared before the Task Force. Speaking at the request of WEDCO, he attacked the housing goals already adopted by the Task Force, arguing that a market for middle- and upper-income housing for families with children did not exist in Queensgate II and maintaining that it was foolish to provide housing for families with children in the absence of very good educational facilities. He then suggested, however, that upper-income housing should go up first, that educational facilities in the area should be upgraded simultaneously, that rental housing should come next, and that all other decisions should be postponed, a strategy not entirely inconsistent with the planning staff's position, despite the antagonistic tone and general direction indicated by his critical remarks.

Puzzled by the apparent contradictions in Taylor's testimony, May prepared a set of revisions to the housing policy guidelines based on the implications of Taylor's attack on the housing goals. This document never went to the Task Force, but May did check by phone with Taylor to see if the revisions reflected Taylor's stand. Taylor said they did. If adopted, these revisions would have postponed the provision of housing for new families with children until the school board improved the quality of the schools. In addition, all the controls to ensure income heterogeneity among families with children and their geographic integration would have been dropped. Put another way, taking this route would have pushed the plan back to a point short of old scheme X, which at least called for the introduction into the area of new families whose socioeconomic characteristics coincided with those of the current population.

For a time it seemed as if the Task Force might adopt Taylor's stance. At an April 16 meeting of that body, Richard Lewis, a member of the Task Force, announced that the West End Community Council delegation to the Task Force had caucused and now wished to table "income and racial heterogeneity" as a "prime goal" and "phase it in later" as opportunities arose. He argued that middle- and upper-income whites would move to such a location only if the Board of Education provided quality education. "Our experience in the West End Task Force," he added, "has demonstrated that these issues will have to be fought on other grounds."

> The harsh realities of American life in this stage of history make the goal of heterogeneity just a dream and a very unstable plank in a planning platform. In spite of these considerations we do have the obligation . . . to achieve decent housing for [the] poor black people of the area. The West End representatives' first obligation is to the people who presently live in the Queensgate II area, and to those who have been forced out by the steady deterioration of the area.

That same day WEDCO presented to the Task Force a list of changes in the housing policy guidelines which moved toward, but did not embrace, the position taken by Taylor. WEDCO's proposals suggested that the housing policy program guideline for a 25-50 percent scheme for income distribution be "considered" rather than "applied" during implementation, and further altered the language of that section to permit income segregation of both families with and without children. WEDCO's proposals also eliminated the requirement for thorough ventilation to permit the construction of back-to-back and abutting apartments and town houses, loosened the stipulations for immediate accessibility to private outdoor living spaces of substantial size for each dwelling unit housing families with children, and opened the door for the building of garden-row structures more than two stories in height, of terrace-type structures more than two stories in height, and of terrace-type structures more than four stories in height. Finally, the proposals loosened the phraseology on staging to allow early construction of low-income housing for families with children.

There matters stood for three months. The planners felt the WEDCO amendments jeopardized the Task Force plan as it had been conceived to this point, but lacked the time and stamina to carry on the fight. The question, moreover, of feasibility as opposed to desirability seemed open to infinite scrutiny and discussion. The planners had been working on the Queensgate II project since October of 1969 under the burden of full-time teaching loads. And the WEDCO amendments offered a glimmer of hope to a variety of interested parties who had been skeptical of the heterogeneity and income mix proposal from the outset, including central business district spokesmen and the city's Urban Development Department, while at the same time appealing strongly to the West End community. Together those forces formed a formidable coalition.

On July 9 the planners formally presented their housing policy guidelines to the Task Force, and on July 16 the Task Force convened to make the decisions. WEDCO presented its amendments. All of them passed. Tom Jenkins rose to make sure that the minutes reflected changes that had been made in the team's recommendations. That was done, and the policy statement was ready for consideration by City Council. On August 31, 1970, it cleared that hurdle.

Given the relaxation of controls in the policy guidelines, it was impossible then to predict what shape Queensgate II would take as the housing policy was carried out. Much of that would depend on the nature of WEDCO's commitment to integration and on the actions of the school board on quality education. But the planning team felt it unlikely that the Task Force's housing goals would be met. If not, the chance provided by Queensgate II to make a dent in the historic pattern of residential segregation by race and income in the inner city would have been missed. Cincinnati, in this view, seemed about to acquire a gilded ghetto.

CHAPTER 7

Research for Planning:

A FRUSTRATED ENDEAVOR

Harry C. Dillingham

After several years it is possible to review the Queensgate II experience with some dispassion. At the time it was occurring this was not possible, at least for me. But even my dispassionate review remains sharply critical of the plan the process produced and of the process itself. In my view, that process produced a disastrously defective plan.

I entered the process in the spring of 1968, when the university's research and planning team first gathered as a group. The charge to the social scientists on the planning team as I understood it was broad—to decide what kind of new area should be developed and to bring our expertise to bear on the problems we anticipated would arise in developing the area. We soon learned, however, that we did not have a free hand to carry out this charge. There were, as one might expect, federal and local guidelines to follow, but there was also a community organization, the West End Task Force, officially recognized by the city, which had veto powers over our plans and which had decreed that we could do no interviewing in the area, not even in the form of casual conversations, without having first obtained permission from the Task Force.

Constraints imposed by the "community" did not end there, however, for we were obliged also to listen to the desires of residents as expressed through the Queensgate II Club. Very early in the process we attended a meeting of that group. Six of its members attended, all of whom urged us to devise a plan keeping current residents of Queensgate II in Queensgate II, where their friends and fond memories were concentrated. They also urged us to make it possible for hundreds and possibly thousands of people who had been forced to relocate earlier to return to the West End by moving into Queensgate II. Most of these people had moved to two neighborhoods, each containing at the time of the relocation small Negro populations (Walnut Hills and Avondale). The arrival of large numbers of Negroes in these neighborhoods was difficult and strongly opposed by most of the white residents, who then fled. The club members of the Queensgate II area were fearful of having to face these same relocation problems, alone and unaided. They said that the present residents had friends and relatives among those displaced and strongly wished for their return so as to reestablish the neighborhood community they had known. They also said that most of the persons displaced from the lower West End by earlier renewal programs wanted to return to Queensgate II.

The Queensgate II Club members also described a classic instance of housing decay in their neighborhood. This was the response of landlords to the impending demolition of their structures in Queensgate II. Knowing that these buildings were to be demolished in the near future, the landlords were not investing in maintenance efforts. The result was an ever-increasing number of vacant housing units and the progressive deterioration of occupied units. In this context, however, the remaining tenants of Queensgate II, *as represented by their neighborhood organization,* did not want to be moved but to be rehoused in better structures.

Thus conditioned, the social scientists held a series of formal and informal meetings among ourselves. We concluded very early not only that each of our academic disciplines had developed some theory and produced some research results pertinent to "urban renewal" or "city planning," but also that such knowledge was often tangential or only indirectly related to the specific problems we faced. My own field of sociology probably had the largest fund of knowledge bearing most directly on our problem, knowledge growing out of studies of urban growth, urban land use distribution, and the invasion of successive ethnic populations into areas, like Queensgate II, adjacent to the central business district. But the construction downtown of residential units on such a large scale was very new. The redevelopment of vast and expensive land areas in or near the central business districts of large cities had never been attempted until federal legislation authorizing and funding it began to be available in the late 1950s. And no such effort had been undertaken in Cincinnati.

After these first meetings we began to reach a consensus on the kind of area we wished to create in Queensgate II, and that consensus defined the situation precisely as the members of the community club and the West End Task Force wished. We agreed that priority should be given to development of the area as a residential one and with full provision for the current residents. Although subsequent knowledge makes this goal seem less than ideal, we accepted the residents' wishes. We knew they could veto our ultimate plans if they wished, and we were moved by sympathy for people whose neighbors had been rudely forced to move away and who, themselves, seemed very frightened at facing the same kind of prospect.

This initial decision helped shape my own research effort, as did the a priori mandate to make the area integrated racially, a mandate which meant that an additional white population would have to be attracted to Queensgate II. Achieving this goal struck me as problematic because the nation's major cities were undergoing massive race riots in this period, and Cincinnati itself had suffered one in 1967. Beyond that, moreover, integrated housing was a very new phenomenon in 1968, and there seemed to be *no* systematic studies of the few instances in which large housing developments had attempted to attract clients of both races. I contacted directors and former directors of several such projects, and was told that they had not been successful. One director's letter was almost tearful. He had undertaken a campaign to enlist white tenants, and a few had come. He described them as militant liberals ready to put their personal convictions to the test. After a few months some began to leave; they had all departed when he wrote to me.

I did find a project in Chicago that had been successfully integrated, however. But the circumstances were so unusual that they could not be duplicated. In that case a hospital had constructed an apartment building adjacent to the hospital. It was surrounded by chain-link fence, and entrance was obtained by showing identification to guards. Most of the tenants were middle-class, childless employees of the hospital. My search for precedents, then, strongly indicated that integrated public housing might not be feasible at that time; and if it were, it could be achieved only by carefully designing and locating the housing.

The proximity of Queensgate II to the central business district was a significant factor in my consideration of the problem. The literature suggested that the attraction and maintenance of a resident population in any particular location must be assumed to result from a balancing of the benefits available in or near that location and the cost of housing in the area. That literature suggested that residential proximity to downtown provided a "least effort" access to certain types of jobs, to a broad variety of goods, and to a broad variety of professional services. But the intense competition for the space in and near the central business district by commercial and manufacturing concerns made such space very valuable, so that its costs per square foot were extremely high compared to locations further removed from downtown. This factor ruled out single-family homes, and meant that land used for residential purposes had to be paid for by high rents per unit of space. The landlords' necessity for large income to pay for the highly valued land could be met, but only if the residents were willing to live very densely in small spaces, such as in multistoried high-rise buildings.

Students of urban housing had also concluded that the amount of residential space people desired was largely a function of the person's family role and age. The decision makers would be adults who had left their own parents' residences. If married, they would fall into one of three categories important for housing preferences: (1) those with no children; (2) those with preadult children living at home; or (3) those whose adult children had left home. People in the first and third categories, as well as single adults, would presumably prefer smaller residential space. This reasoning about housing preferences suggested that the most viable target populations for location in Queensgate II would be single adults and married couples without children.

I based these conclusions on broadly measured historical trends and believed them to be subject to some modification, because the structure and growth patterns of major cities had begun to change. Large urban centers such as the Cincinnati metropolitan area had continued to grow, but not so much by increases in the population *within* the core cities' legal boundaries. Indeed, this population had begun to decline after 1960, although the evidence for this decline was still scant in 1968. Nonetheless, it seemed obvious that suburbanization had set in with a vengeance after 1950 and that it was not merely a residential phenomenon. Because of high taxes, cost of maintenance, cost of land for expansion, and traffic congestion, many industries and commercial activities also began locating outside of cities' legal limits (they were both following *and* attracting residents to the suburbs). The only business (industrial or commercial) that did not show long-time declines in the central business district and the central city as a whole were offices employing literate clerks, usually females, who were traveling further to work as the center of gravity of residence shifted away from the city.

In light of the above trends and forces it seemed probable that our redevelopment would be most *viable* if we constructed multiple-story apartment buildings with small units to house one or two adults. These would probably be young adults, and most often young women, employed as clerical office workers in the central business district. Other types of adults who were known to have some attraction to this location and type of dwelling were elderly retired couples or widowed or divorced persons. All of these people tended to avoid problems associated with dwelling maintenance, whether in yard grooming or meal preparation. The elderly lacked the energy, while the young, whether childless-married or single, preferred to spend their prodigious stores in other activities. In addition, neither category needed or desired large dwelling units, and both found a location near downtown advantageous. The young would have access to a concentration of entertainment and thereby to other young people with similar tastes, and the elderly would have access to the widest array of shopping goods and services (including medical) in the urban milieu.

As my contribution to the research effort I proposed to test the applicability in Cincinnati of these generalizations about downtown housing and target populations by conducting a telephone survey of the metropolitan population to determine more firmly and precisely *who* we could expect to attract into a redeveloped Queensgate II. A small budget for research severely restricted the size of the sample, which was drawn from the Cincinnati and suburban directories by selecting 294 people whose business addresses lay within the central business district. A commercial firm conducted the interviews, questioning women only (unless the respondent was an unmarried male), because prior research showed the wife's decision to be most important in selecting the family's place of residence.

Survey results showed that the most important factors influencing the choice of residence near the central business district were age, marital status, children living at home, and time spent in travel to work (not occupation or education). Those showing *any* positive interest in living near downtown were young (50 percent of those under 25 had some interest), or older (40 percent of those over 55), but older people predominated among those with the *most* positive interest. Single persons were more interested than married persons, and the *most* interested singles were divorced, separated, or widowed. Among previously married persons, those with no children or just one child living at home dominated. Over 50 percent of the sample living within 4 miles of downtown and 40 percent living over 8 miles away expressed an interest in living near downtown, while the figure for those located at the intermediate distance fell below these levels. Time spent traveling to work (as opposed to distance) also proved crucial, for the longer people who were disposed to live downtown traveled to work, the more intense their desire to live in the central business district.

We also asked people how they felt about living in an integrated neighborhood. Of the white respondents, 31 percent said they were unwilling to live in such a setting (but I suspect the percentage would have been higher if we had asked about living in an integrated apartment building). Those opposed to living in an integrated neighborhood were most likely to be married homeowners with children living at home.

Most of these findings were not surprising. We did expect, however, that the percentage of singles who had never been married who would be interested in living near downtown would be higher than it was. We were also surprised that the level and intensity of interest in such a location among the divorced, separated, and widowed. But these findings and the other data suggested that it would be very difficult to attract to our neighborhood married people with children.

Finally, the sample contained too few Negroes to analyze separately with any confidence. But research by others after our study suggested that they might harbor the same housing preferences as whites. That research draws an analogue between the earlier white immigrant experience and the experience of black migrants to American cities. That is, immigrants arriving in our largest cities without money or command of the English language, and with access only to the lowest-paying occupations, *had to* settle in housing near central business districts, irrespective of their family responsibilities. As they prospered, however, they moved away. After the 1950s, Negroes, too, were increasingly found moving from inner cities to suburban areas, and their numbers there increased at rates equal to that of whites. At the same time, the rate of growth in the number of Negroes *within* core cities remained stable or declined.

Our team accumulated additional data on housing preferences and socioeconomic and family status, however. The psychologist in our group, Dr.

Harold Fishbein, designed and collaborated with various other planning group members in interview studies of residents of Queensgate II, residents of adjacent public housing projects, and residents of neighborhoods into which former West End residents had relocated. He found that the families and single persons who lived in Queensgate II were long-time residents (median of 18 years) and middle aged or older. Half the households did not contain an employed person. In addition, 70 percent were childless, and about half of these were unmarried persons living alone. Of those with jobs, only 12 percent worked in the central business district or an area adjacent to it, and relatively few of these people reported shopping downtown. More importantly, only 40 percent saw the area as suitable for rearing children, and 75 percent said they would move to another area for better housing at the same rent. Indeed, 50 percent of them had recently considered moving away.

Another survey supervised by Dr. Fishbein interviewed a sample of residents in three neighborhoods to which many, if not most, of the former residents of the renewed portion of the West End had moved. These persons were asked if they would move to the general area of Queensgate II "for better housing at the same rent you are paying now?" There was no relationship between previous residence in the West End and an affirmative answer, a finding suggesting that Queensgate II Club leaders had seriously erred in asserting that these former residents would especially like to return to the Queensgate II area.

We now had considerable information bearing on the question of what to do with Queensgate II. Studies of land use distribution, land values, and migrant succession in large American cities uniformly showed migrants accepting densely populated inner-city quarters only until they or their children had assimilated, at which time they moved outward to areas offering more amenities for forming families and rearing children. Those and other studies indicated that businesses most resistant to suburban location were service sector businesses with labor forces consisting of well-educated professional, managerial, and clerical workers (my own survey found that persons working in Cincinnati's downtown were highly educated—50 percent of them had *more* than a high school education and 75 percent had a high school education). My survey of housing preferences found that persons with little or no family responsibilities—specifically the childless young and the elderly childless—were most likely to want to live adjacent to the central business district and that those living farthest away and spending the most time in travel to the central business district were also likely to be interested in relocating near the central business district. Dr. Fishbein's survey disclosed a *resident* population of elderly, childless couples with a high unemployment rate and with few in the work force holding jobs downtown, and most wanted to move or were considering moving elsewhere for improved housing.

What did this data suggest for the renewal of Queensgate II? First, only a certain kind of business might locate on such a site. Since retailers, wholesalers, and manufacturers had been leaving large, older central business districts, including Cincinnati's, they should not be encouraged to locate in Queensgate II. Instead, offices of various kinds should be sought.

For residential development, two target populations seemed most salient. One of these was people who might be attracted to the area because it would reduce time and cost of transportation to their jobs and also provide easy access to an abundance of personal services. The most likely contingent of these people would be young, married couples without children and older couples whose children had left home. These people had higher than average educations. They were predominantly white-collar workers. Their housing standards and the rents they would pay for such standards were above average. Moreover, occupancy in existing high-rise apartments in the city suggested that apartment living was not repulsive to the group as a whole.

Current residents constituted the other likely target population for residence in Queensgate II. The great majority of them were quite similar to the older segment of the first target population, that is, middle aged and above with no children living at home (though they differed by race, education, and level of unemployment). Since their incomes were low, and with half of them on some form of government-provided support, they could pay, or could be assisted to pay, for much less expensive housing. Therefore such housing would have to be designed for high density, as in high-rise apartments, where maximum use of the ground space would be achieved with minimum expense. And provision would have to be made elsewhere in the city for more spacious homes for the 30 percent of Queensgate II residents with children.

Since there were 117 acres in Queensgate II it seemed obvious that provision could be made for one or even several service-type office buildings, for buildings to house persons who worked downtown, or for buildings to house current residents. More importantly, provision could be made for all three or for any two of these kinds of developments, but I did not think it made sense to recommend the construction of housing for current residents in the new Queensgate II.

The current residents did not work in or near the central business district and did not possess skills appropriate for most downtown occupations (white collar). They had located in Queensgate II before the 1960s because it was the largest Negro ghetto in Cincinnati, which is to say, the only area where they were tolerated. In the later 1960s, however, the toleration and legal support for these people to live elsewhere in the city was sufficient to give them easy access to several other neighborhoods, and access with varying degrees of opposition in any neighborhood. Housing could be erected in such outlying areas much more cheaply than in Queensgate II. Indeed, more spacious housing (more rooms and more yard space) could be constructed in such outlying areas, because the land was not sought for more profitable, intensive uses, such as office buildings and high-density, multi-story, high-rent residences. In sum, it seemed to me inefficient and unprofitable to erect new residences for the current Queensgate II population in Queensgate II.

The planning team as a whole disagreed with me, however. Influenced by the Queensgate II Club and its parent organization, the West End Community Council, the team recommended that *priority* be given to housing for the current residents of Queensgate II. The team also acceded to these organizations' insistence that those persons previously removed from the redeveloped lower West End be given a secondary priority. Nor was that all. The planning team also recommended the construction of 1000 units for families with children, some of which would contain three *or more* bedrooms. This meant there would be an enormous increase in the child population of an area in which elementary schools were already packed, a catastrophic move unless new schools were erected before or concurrently with the repopulation of the area. Yet the educational plan was not completed until after our team disbanded, and, when finished, called for public educational expenditures that could only be called phantasmagorical in light of the financially strapped condition of the Cincinnati public school system and the growing reluctance of Cincinnati school district voters to pass school tax levies.

Given the process by which the team developed its plan, it is not surprising that the final product should have been so unrealistic. The influential role played by the two West End neighborhood organizations helps account for some of the plan's defects, although it should be remembered that the team could have listened to these groups without becoming sympathetic to their views on the desirable nature of the plan. And, as a social scientist and, I like to think, a sensible citizen, I am certainly not going to fault the assembling of a research team to assist in the planning. After all, it seems obvious to me and, I trust, to the reader that the expenditure of an estimated $12 million in an urban renewal project requires the careful analysis of the forces (usually economic) that have produced and are shaping our neighborhoods and cities. But that analysis must be thoroughly and reflectively done, virtually an impossibility in the Queensgate II case because of the miniscule budgets available to our researchers and because of the tight deadlines imposed upon them. Moreover, that kind of analysis simply cannot be done when the researchers must in a short time make both policy decisions and technical judgments, as we had to, about which goals are feasible and how much they might cost. As a result, we had to spend interminable hours in *deciding our policies* during the summer of 1968, the only time we had in which to conduct the investigations—such as surveys of people, legislation, and the social science and historical literature—the results of which would have to be known *before* policy decisions could be debated intelligently.

For all these reasons, the Queensgate II planning team produced a defective plan. If the plan is carried out, I believe it will enlarge this ghetto. The geographic distribution of the planned buildings does not maximize the separation between units designed for persons of different social classes or for persons in different stages of the family cycle (before, during, and after children are living at home). Those living elsewhere, employed, and with children, will be least attracted to such an area. And if most of the residents as of 1968 are provided cheap housing in the new Queensgate II, the area will continue to have high levels of unemployment and poverty. The only persons with children living at home who can be attracted to Queensgate II in any numbers will be ones who are provided with government assistance. Not only will the ghetto be expanded, but this valuable land will not be used to enhance the economic viability of the central business district or to maximize the potential tax returns to Cincinnati—quite the contrary.

CHAPTER 8

Research and Planning for Education

Lewis A. Bayles

I reacted ambivalently when I received an invitation, in the summer of 1968, to join the Queensgate II project. I had no previous training or experience in city planning. My background in education was in training teachers in the philosophy, sociology, and politics of education, and I was therefore ignorant of the details of school building. Moreover, I had been at the University of Cincinnati for only one year, had only superficial familiarity with the city, and little contact with or access to its decision-making processes. And I was skeptical of the utility of the "outside expert" in planning on the grounds that those who carried out plans should play the key roles in formulating the plans.

Yet I was attracted by the challenge to see school planning, indeed, education itself, in an unconventional way. Ordinarily, communities grow and school buildings and programs are designed only after considerable lag. This project, however, proposed the insertion of educational factors in social planning at the start, and raised the possibility of using schools as an instrument in *building* a community. I also believed that universities and colleges of education, particularly those in urban areas, had obligations to the local community, and participation in the project seemed a way in which I, as a faculty member, might contribute. I certainly believed that I could learn by joining the project and I felt that if I became a liability to the planning effort I would realize it and gracefully retire.

I remained uncertain for some time, however, and I finally talked over my reservations with Tom Jenkins late in the summer of 1968. He indicated that my ignorance of planning was not much of a handicap because there was little precedent in city planning for what my job would be. I would be a member of a faculty team, he explained, and so involved not just with schools but with the educational impact of all aspects of the plan. He added that educational planning would move only to the point of proposing programs and that school officials and facilities experts would be involved when necessary in the planning process. What was needed, he stressed, was a broadly knowledgeable and possibly creative bridge among the many specialists. My job would be to provide options and alternatives and to project their likely consequences, not to make final decisions.

After this discussion, I decided that I was as well qualified for this novel assignment as anyone else I knew. As a generalist, I was aware of trends and developments at many different educational levels and in many educational specialties. My background in educational philosophy prepared me to avoid a prior commitment to one program or one philosophy of education. This training also helped me to distinguish long-run developments from superficial fads, a confusion which sometimes afflicts educational thinking. And my reading in educational sociology led me to view education broadly as a social process, something that occurred partly in school and partly outside, and that therefore could not be isolated from other social processes. I needed to be aware not only of what was known about relations among schools and communities but also of those beliefs that might affect policy and that were based on conjecture or convention.

Finally, it seemed that my status as an outsider relative to the Cincinnati school system was perhaps an asset. Most informed people I knew believed that the Cincinnati system was as centralized and bureaucratic as the typical urban school system. Even the most thoughtful, capable, and creative individuals from within the system were very reticent even in imagination to venture beyond their own functional divisions or to elaborate new ideas without prior approval from superiors. Given all this, an outsider, unfettered by organizational divisions and chains of command, might serve as a catalyst in a new mix of intellectual and social forces. Feeling that I had some assets and that I was sufficiently aware of my shortcomings that they could not harm the project, I agreed to join the university planning team.

The first step in the planning process was to describe the educational situation in Queensgate II, discover what plans were already envisioned by the school system, and to assess the educational needs of the current population. For the last I relied heavily on a Model Cities study and on studies prepared by the West End Special Services project. For general information on the Cincinnati public schools system I turned to an extensive school survey being completed by a special staff organized by the Midwest Administration Center of the University of Chicago, a rather conventional survey too broad to provide much detail about any one school or area, but containing the facts necessary to build an overall perspective and information on some specific problems, such as the lack of master building plans. In addition, I conducted interviews with both school and community leaders, the most productive of which were rather long ones with the principals of the three schools directly involved, Hayes Elementary School, Porter Junior High School, and Taft High School.

From these preliminary inquiries I emerged with some convictions which, though strong, could not be easily given the solidly factual basis that might make them persuasive to others. The first was that educational planning for Queensgate II should reach beyond the physical boundaries encompassed in the original contract. Within the physical boundaries of Queensgate II was one school site containing Hayes Elementary School, with a capacity of 900 kindergarten through sixth-grade pupils, and Porter Junior High School, with a capacity of 700 seventh- through ninth-grade pupils. Across Lincoln Park Drive, directly adjacent to but outside the project boundaries, was Robert A. Taft High School, with a capacity of 1100 pupils from tenth- through twelfth-grade. Each of these schools drew students from an area larger than Queensgate II. While pupils from the Queensgate II project area constituted a significant *minority* of the pupils attending Hayes Elementary School, Porter's district boundaries included nearly one-third of the basin pupil population and Taft High School boundaries were virtually congruent with the basin limits.

I also concluded that the programs and services available in these schools were largely determined by the school system's central administrative office and the Cincinnati Board of Education, without significant community participation. Curriculum, budget, and personnel policies were largely city-wide, and special services, such as health, visiting teacher, and special education programs, were provided by a variety of agencies at many different points in the city. These circumstances suggested the utility of a close interrelation of Queensgate II both with the schools of the basin and those in the rest of the city. Anything affecting the schools serving Queensgate II had to affect many others, while developments outside the area necessarily affected the educational facilities and services available to Queensgate II residents. Educationally there seemed no way to insulate Queensgate II from the problems and prospects of the basin and ultimately the city as a whole.

Having reached these conclusions, I inquired about plans the school system might have that would affect Queensgate II. There were very few, and other changes that might affect the schools of that area were not of high priority. This attitude was exemplified in my early conversations with Guy Buddemier, an assistant superintendent in charge of the Division of Research Statistics and Information and also the representative of the school system on the West End Task Force. As the school system representative, he felt he had nothing to say regarding housing, social services, recreation, or other matters not directly affecting the schools. If something came up that directly affected the schools he said he would refer it for study to the appropriate school division and report back to the Task Force. He saw the situation largely in terms of buildings, and since the local school population had declined below the rated "capacity" of the buildings, even a substantial increase in pupil population could be accommodated by using the buildings to capacity, by revising attendance area boundaries, and perhaps, if necessary, by altering uses of current buildings. He also pointed out that Hayes, Porter, and Taft schools were all built in the early 1950s and were physically adequate if architectually conventional, while the system contained many older schools, some of which were overcrowded, and some of which were housed in temporary facilities. For these reasons, school authorities put their highest building priorities outside Queensgate II.

I also think they were cautious and passive about our project because of their past experiences with urban renewal. Earlier such projects in Cincinnati had used school building expenditures as a part of the necessary local contributions but usually provided no federal renewal funds to help cover the cost of educational improvements. Thus, the school system had contributed to urban renewal but had received nothing directly in return, a condition school officials did not expect to change.

The tight financial straits of the school system added to the cautious mood of local educators. A school bond levy had recently failed, and there seemed to be little support in the city for property tax increases or in the state for changes in tax policy. The resulting austerity meant that few resources were available for anything but immediate and urgent needs. Indeed, it was customary that no proposal for increased expenditure would be considered that was *not* tied to a source of increased income, but proposals that did not require funding or that had external funding seemed to be adopted with relatively little scrutiny. Thus, the school system operated a wide variety of surprisingly controversial programs using outside, mainly federal, funds. This situation left the impression that new ideas might be accepted, provided they paid for themselves.

Another impression that emerged from my preliminary explorations dealt with the relationship of schools and the other "needs" of the community. As I examined the many reports and studies detailing the problems of Queensgate II, I was struck by how frequently proposals concerning health, welfare, social services, integration, or employment involved the schools or schooling. I was skeptical of much of the enthusiasm for the notion that schooling might provide a panacea for social ills, but it was clear to me and to others that the belief was widely held that schooling should be involved in any plan for community improvement.

I also reached some early conclusions about the politics of planning for Queensgate II. I entered the project convinced that interested groups had to have sufficient influence upon the decision makers in the schools to assure the implementation of the plan, to carry through on the program implications, and to make adjustments in the plan required by unanticipated changes without losing the essential thrust of the plan's ideas. I was intrigued by the possibility that the West End Task Force might provide a solution to these potential problems. Even though its role was purely advisory, it did seem to have considerable influence, for it contained representatives of both downtown business interests and authentic community interests. The power of the business interests combined with the sensitivity of the grassroots leaders could, in theory at least, provide a power base for continuing productive influence upon school decision makers. The problem was to devise a plan that would gain Task Force support and build into it the mechanisms for continuing access to decision making for the Task Force.

We decided to postpone educational planning, however, until the West End Task Force had made some basic decisions on housing, residential density, and population mix. The university team decided to try to eliminate only those options which were clearly impossible and to reduce the remaining options to as few as possible, describe their most likely consequences, and leave the basic choices among these options to the West End Task Force.

This procedure meant that school improvements would have to be examined in two ways, both as means and as ends. What educational facilities and programs would be necessary to serve and/or attract the projected population? What impact upon educational goals would be likely to result from the projected populations?

The university team came up with three alternatives (labeled X, Y, and Z) for housing, residential density, and population mix, none of which I viewed enthusiastically. Two alternatives (X and Y) provided only very small increases in housing for families with children, increases that exactly matched the numbers the school system could easily have accommodated in existing facilities. But the small number of children, even if they came from strongly school-supportive families, would not have been sufficient to alter the predominantly disadvantaged character of the students. It was also quite possible that strongly school-oriented families would not send their children to public schools, and that the small proportion of families with children would not be large enough to increase the political influence of the community upon the school system, including voter decisions on school bond levies.

The third alternative (Z) seemed particularly dangerous from an educational point of view, for it contemplated increasing the population of young adults and the elderly. Many studies of voter behavior in school board elections had indicated that young adults, because of their poor registration and voting habits, and persons beyond the child-rearing age, because of their desire to keep tax rates low, tend to provide little voting support for public schools.

Despite my reservations about these housing alternatives, I began to think about ways to improve the schools in Queensgate II without adding greatly to their cost. This constraint, plus the general desire that schools help meet a variety of community needs, led me to consider the highly flexible "community school" idea, under which schools would be vocational, social, and social service centers rather than merely educational facilities for the standard school-age population. The idea had several attractive features for the Queensgate II project. The waste inherent in leaving buildings unused for large periods of time was obvious, and using school space or space close to schools for social service agencies would make the services more easily accessible to children and their parents. In addition, such schools could be opened in the evenings for adult educational programs sponsored either by the schools or social agencies, again at a minimum additional cost. To counter the fear produced by lack of lights on the streets and around the school, we proposed improved lighting for the school yard and the streets around it, a step which would also make the Porter-Hayes play area available for use after dark.

We also proposed more adequate recreation facilities, a suggestion which received support from many quarters, especially among school people and youth-oriented community leaders. According to our version of the community school idea, gymnasium facilities would be available for little added cost. It also called for space for a playing field suitable for football and baseball and a stadium for Taft High School, the only high school in the city without such a facility.

Another element of this proposal concerned an area technically outside our boundaries. There had been in Cincinnati for some time the notion of decentralizing vocational education programs, which had in the past been largely handled at Courter Technical High School, by adding more vocational facilities at Taft High School. This seemed desirable for Queensgate II. It would provide for a greater variety of programs and would also provide facilities for increased vocational training for adults. It also reinforced the community school idea and seemed likely to draw on state and federal funds.

If these improvements could be made, it seemed to me, the Hayes, Porter, and Taft schools could serve as the center of community activity and interest. All sizes and ages of people at all times and all seasons would be moving through and using the facilities. Since the area served would be larger than Queensgate II, the new facilities would establish one of the linkages urged by the overall planning goals. The schools which in the 1950s and 1960s formed barriers isolating Queensgate II would, when reorganized, help reintegrate it with the larger community.

These proposals constituted minimum proposals, however, and they were largely remediative and designed to provide an education in Queensgate II comparable to that available elsewhere in the city. They would improve, but not significantly alter, basin schools or schooling. Therefore, we designed another set of proposals to stimulate comprehensive, dramatic, and visible change. This we called an educational park or complex.

Many later felt that the educational complex was designed for the primary purpose of making middle-income family housing marketable. This, however, was not the case. The notion of applying the educational park concept first occurred to me during one of my early walking tours of the area and well before we discussed alternative population mixes.

Specifically, the possibility of an educational park first presented itself as I stood on Lincoln Park Drive in front of Taft High School and tried to imagine the area with all its dilapidated buildings removed. Due east along the Drive stood Music Hall, old and ugly, but still in use as a performing arts center for the city. At the west end of the Drive the school system had taken over the former railway terminal and was developing it as a science center for the entire city. At the apex of a triangle I could visualize (but not see) City Hall and St. Peter in Chains Cathedral. Using these buildings in an outreaching "city as a schoolhouse" program (something like this had been done in Philadelphia) could provide an educational program that could not be duplicated anywhere in the city. And increasing the conventional school space by using these buildings would expedite the closing of those ugly, poorly designed, and superannuated schools still in use around the central city. Finally, and somewhat later, I added to the park proposal facilities for programs for retarded, hearing impaired, and emotionally disturbed children who were being housed "temporarily" in old "underused" schools close to downtown that were designed with no attention to their students' special needs.

The idea of such a park in a compact and contiguous area was quickly rejected. To concentrate all the necessary facilities and open spaces within the Queensgate II boundaries would virtually eliminate any other land use, including the housing for the current residents that was such a firm goal for the project. The general objective, however, drew some favorable response, and the scheme was revised into a cluster arrangement which included several elements. It called for a comprehensive high school with programs for a student population of at least 2000 pupils, the standard minimum number for a comprehensive high school. In addition, a significant proportion of new space would include facilities for vocational education which were already projected and to be financed with local and some state funds. Our proposal to use the science center and the performing arts center would mean that the conventional classroom space already at Taft would suffice for a substantial increase of students and could be used in the afternoons and evenings to upgrade adult employment opportunities. In this way a facility to meet the needs of both the college-oriented and the job-oriented could be met with little added cost. Average students would benefit by being able to prepare for post-high school employment, but still take enough academic courses that they need not give up the chance to attend college.

Slanted in the same direction was a proposal for a community college under the shared control of the Cincinnati Board of Education and the University of Cincinnati. This was an idea that had been tossed about in Cincinnati for some years. Bringing low-cost college programs closer to downtown was appealing to basin residents, but would also be of value to the college-oriented middle class. Instruction would take place in conventional classrooms during late afternoons and evenings, when the buildings would be open for adult and recreational programs. Some job-oriented programs could emerge from this mix. The community college would also tend to shift the political balance by involving the university, with its distinct power bases, in the continuing development of the basin. In addition, stadium facilities were easily justifiable under this plan and could benefit the whole area aesthetically by visually opening this largely compacted district.

The major new school construction envisaged in the educational cluster complex was at the elementary and junior high school levels. The primary objective here was to work toward total replacement of the superannuated conventional facilities (other than Porter and Hayes) still being used in the basin. Modern school construction stressing open spaces and multiple uses would have had many consequences. At the minimum it would provide pleasant and decent places in which children could learn. The visual effect could also appeal to middle-class parents, as well as improve the appearance of approaches to the central business district.

Another important part of the cluster concept had to do with schools for exceptional children—the mentally retarded, visually and hearing impaired, emotionally disturbed, and orthopedically handicapped. Such facilities should be in central locations with excellent transportation, could share many specialized personnel, and could benefit from the easy availability of regular school facilities, where exceptional children could be brought back to the regular classroom. The specialists who worked in the schools would be able to live close to their work. In addition, such schools would offer a number of paraprofessional and lay positions, which Queensgate II area residents, particularly mothers with children, could effectively and conveniently fill. Finally, special education programs had received more support from taxpayers than had regular schools.

We did not develop a program plan for the educational complex. We tried to do just two things: arrange (1) those conditions which would make the complex an attractive place to work, and (2) those which would focus outside attention on program planning. We expected the place to attract creative teachers and administrators who would insist on getting strong support facilities and who might even secure pleasant nearby places to live. Interest from downtown business and civic leaders and university schools of education would stimulate demonstration projects which in turn would generate publicity and additional interest. If each of these various groups participated in the program decision-making process, the outcome could not be predicted but the process might stimulate and invigorate the conventional bureaucratic style and attract those professionals in the school system to whom change was a challenge, not a threat.

When originally proposed, the educational cluster did not engender much enthusiasm. School people tended to agree about its basic soundness, but viewed it as impossibly idealistic for this public education system. Community reactions were favorable to the plan, but doubtful about its being adopted by the school system. Even I was ready, by the close of 1969, to give up the idea as one whose time had not yet come. I felt that only a significant increase in pupil population would stimulate more than a few of the school decision makers to such a dramatic and seemingly daring innovation. Yet the housing alternatives did not go far enough. Of the originally projected alternatives, only Y included any substantial increase in the number of families with school-age children, and not by my estimate sufficient to obliterate the low status image of the schools even if the children came mostly from strongly school-supportive families. It seemed, however, enough to at least sustain demand for college preparatory-type programs in the area. This I felt was necessary in order to preserve a comprehensive program to the benefit of that group which could sustain academic capabilities and interests in spite of a discouraging environment.

I was about to drop the educational complex idea when I heard the reactions of the outside economic consultants to our preliminary housing alternatives. The economists doubted that many middle- or higher-income families with children would be enticed to move to Queensgate II unless the "amenities," including schooling, were clearly committed *before* development began. Unfortunately, no more room for such improvement was available within the project boundaries, and the same report indicated that a lower concentration of population was also required. Nonetheless, the economic consultants' report, more than any one thing, provided us the motivation to retain and develop the cluster idea, but as an addendum rather than as part of the plan itself.

We had concluded that the housing could not be built or even planned until the school system had definitely committed itself to the school improvements. We believed that the school system customarily made commitments for populations that existed, not for those merely planned or projected. Therefore, the basic planning strategy was to continue to develop the cluster proposal and attempt to gain commitment to it from the school system. To do so, we felt the proposal had to be concrete and specific in order to evoke definite reactions, but sufficiently open to provide flexibility as detailed planning took place in the process of translating the plan from paper to bricks and mortar. At this point I tended to resist proposing concrete details, even though I suspected the project staff may have felt I was stalling. Two factors, however, could not be evaded. For me to submit detailed plans required time and effort from school system experts in both building and programs, few of whom could be convinced until close to the end of the project that this was more than a futile intellectual exercise. Second, I was, through experience, fully aware of the bureaucratic infighting tactic of exploiting a minor fault in a detailed proposal to destroy a much larger idea. In this case, I feared the use of quibbling over concrete details to kill the educational complex idea.

In any case, these constraints guided the later stages of the education planning for Queensgate II. Preliminary proposals were submitted to a variety of agencies and groups for their reactions and suggestions, and very few changes were suggested or adopted. Proposals regarding programs were gradually elaborated as I became more aware of programs that were close to implementation and which seemed beneficial to the area. Most agencies liked the ideas but tended to condition any real commitment upon the previous commitment of others, and seemed unwilling to take the first step of committing resources sufficient for serious study. This made the process slow but tolerable because I knew the elements of the plan were closely interconnected and required coordination that cut across traditional bureaucratic divisions. The project staff, however, was anxious to complete the report and fulfill the conditions of the contract, and the full education plan, including the educational complex addendum, was turned in after the Queensgate II plan.

As a consequence of this time problem, the Queensgate II plan contained two features, the Town Center and satellite schools, which came from the project staff and which contradicted our previous and more ambitious thinking about education. The Town Center, which was publicized before I could react to it, contained a variety of community facilities and agency branch offices and was to function as a focal point of community pride, interest, and interaction. This, of course, is exactly the role that the community school concept proposes that schools should serve. What reasons motivated the decision to place these facilities outside the school I was never able to determine, and I could not get the decision reconsidered.

The satellite school moved from initial suggestion to incorporation into final plans with remarkable rapidity, although in this case I was present when the suggestion was first made. The idea was to provide within each unit of family housing space for an ungraded open plan for preschool and primary grades. This idea was originally mentioned by the man appointed by the school system to serve as liaison to the project. Although the planners on the university team liked it, my original reaction was tepid. The scheme entailed preschool, ungraded instruction with an open classroom orientation, widely accepted ways to improve schooling for disadvantaged children which also appealed to many middle-class parents. On the other hand, it also seemed to provide a protected environment for the privileged few who lived in Queensgate II, whereas improvements at Hayes would benefit many inside and outside the immediate area. I wanted time to weigh the idea, but before our next meeting housing plans had been drawn using all physical space available in Queensgate II and requests were being made for specific space-per-child estimates. Guidelines to limit the possibility of resegregation and to assure priority for the present residents as well as mandates for community advisory councils were also included, but I still had many misgivings.

It was the satellite schools, however, that came to be the symbol of the Queensgate II educational recommendations. No other educational proposal received more publicity in the newspapers or occupied more space in the final report. It is an interesting commentary on the planning process that the proposal which had the least to do with improving schooling in the inner city was the one most quickly adopted and widely noted.

In any case, it is very difficult for me to assess the Queensgate II experience. There seemed to be at one time an opportunity, through the educational complex idea, to dramatically improve educational opportunity throughout the basin. A similar opportunity may never arise again. If it does, the project recommendations are on record in the addendum to the plan and may be resurrected at a later time. It may at some later date be seen as an idea for its time.

The educational proposals that were finally made came out as "recommendations to related agencies," when at one time it seemed that the schools would be the central theme. Of those proposals, the one that seemed most likely to be adopted (the satellite school) was one whose effect would be felt by the smallest number. Other parts of the plan, such as recreation and social services, were better perhaps than they might have been otherwise because they met needs that we analyzed as emerging from educational problems. Without consideration from the educators' viewpoint they might have been different. But the final assessment of the Queensgate II experience is for time and people other than the participants to make.

CHAPTER 9

Managing a University Team in Partnership Planning

Thomas H. Jenkins
Jayanta Chatterjee

Although the university was one of the partners in the Queensgate II planning process, it, like the other partners, was not a monolith. To handle the work, the university broke into essentially three components: the university team's planning staff; its social science research and consultant group; and the university administration. Each of these components was supposed to have its own separate function. The planning staff, composed of faculty and students in planning and architecture, was supposed to do the planning and design work. The faculty social science research and consultants group was supposed to develop social scientific guidelines for and analytic responses to planning and design proposals. The university administration was supposed to oversee the city-university contract by attending to the formal and official relations between the two contract parties.

As director and assistant director of the university's planning team, we hoped not only to help devise a plan acceptable to the West End Task Force and the city but also to help point urban renewal in a new direction. We wanted to influence the process enough to produce a plan different in content

179

from those recently adopted in Cincinnati and most other cities. We hoped to make citizen participation a meaningful rather than a token part of the planning process. And we hoped to make social science research an integral part of the planning process, particularly within the context of a team planning effort, rather than something that occurred before or after the making of the plan. It seemed to us and others at the time that the unique three-way partnership of university, city, and community provided a means through which all these things might be acomplished.

On the whole we think it worked, despite a series of difficulties involving or confronting the university team which disrupted the process and which more than once threatened to dissolve the process. Some of the difficulties were unprecedented in the experience of the participants, while other more predictable and familiar ones looked different in the context of partnership planning. Some difficulties involved conflicts between various elements within the university. Some came from conflicts pitting the University planning team, or parts of it, against the representatives of the city or of the West End black community. Some stemmed from conflicts placing the university planning team in a crossfire between contending forces outside the university. And some occurred within the university planning team itself and arose from the attempt to integrate social science research into the planning process.

Organizers of the university planning team first faced the problem of finding university faculty members with the time, credentials, appropriate academic specialty, and inclination to engage in a novel, complicated, and concentrated one-year planning effort, during which they would be expected to carry out teaching responsibilities. Planners and architects from the College of Design, Architecture, and Art came aboard first and expressed few qualms about participating in the project. They were used to teaching while working on "real-world" problems as consultants and had experience both in the "real world" and in the college with team efforts projected along a sequence of deadlines under tight time constraints. Social scientists from the McMicken College of Arts and Sciences proved more reluctant, however, and for a variety of reasons unrelated to the college or this particular project.

Like most social scientists, those in the College of Arts and Sciences had little experience working on "real-world" problems, either as individuals or as members of a team. Few had ever participated in team research efforts even on the kinds of problems normally pursued by academic social scientists. Almost none of them felt comfortable about the prospect of doing research that had to produce a particular kind of product at a particular time through a particular sequence of decisions on aspects of the general problem. And most expressed reservations if not downright opposition to the very idea of doing social science research on problems that they themselves could not fully define and that might not contribute to the solution of problems and the advancement of theory in their respective academic disciplines.

Social scientists in the College of Arts and Sciences at the University of Cincinnati, in short, were like most social scientists in other academic institutions. That meant that most of those we got were people for whom not only research for planning in a partnership arrangement but research directly integrated into any kind of real-world planning process would be strange. It also meant that few were willing to participate, which limited the pool of talent out of which the organizers of the team made their selections. So, while we got social scientists from disciplines we wanted, not all of them commanded the appropriate specialty within their discipline. To be sure, our historian, geographer, and educator were all urbanists, and therefore brought with them credentials suitable for our task. But we had a regional economist with experience in housing as a national issue when we really needed a real estate or housing economist with expertise on the question of race and the problems of housing as a neighborhood and city issue. We had an experimental psychologist when we really needed a social psychologist with experience in conducting attitudinal research through survey techniques. We had a political scientist who specialized in public administration when we really needed one working on community politics and especially the politics of the black militant movement. And we had a sociologist with a specialty in social organization when we really needed one with a strong background in human ecology or migration studies.

Despite these deficiencies in the credentials of some of the social scientists, they all produced ideas and research useful to the team as a whole. Indeed, some of these contributions struck us and some others as quite impressive. The list includes Lewis Bayles's proposal for an educational complex, which is described elsewhere (Chapter 8) and on which several members of the team worked. In addition, geographer David Ames devised a new "adaptive" planning scheme for the use of a fringe area of existing stores around the project site as a staging ground within which to help relocate, sustain, or phase out small businesses in a gradual and humane manner over an extended period and through the private market. In addition, Donald Heisel, our political scientist, headed up a study of job opportunities for

West End residents which linked the general land use policies of the city to industrial development policies for an area adjacent to Queensgate II, tied that to policies on transit routes and fares to the West End, and prescribed hiring, manpower development, and welfare policies for the city and public agencies.

One of the most impressive and certainly one of the most influential pieces of policy research was urban historian Zane Miller's study of the Queensgate II area's history. Though a history, it can be called policy research because its assessments of what transformed Queensgate II from a lively to a dying neighborhood inspired the basic physical and social planning policies built into the plan; namely, that the Queensgate II area should become an identifiable residential area reconnected physically and socially to its surrounding area and to other parts of the city. And the adoption of those policies by the city in effect reversed the city's previous policy of expanding the central business district westward and of replacing what might remain of West End black residences around the new downtown with non-central business district types of industrial and business developments. To put it bluntly, adoption of those policies placed the city on record against Negro removal and in favor of residential development for a heterogeneous population around if not in the central business district.

As impressive and as important as these and other studies were, however, we nonetheless encountered difficulties in working with the social scientists. For example, some of the architects and planners, who, like their colleagues elsewhere, were more practiced at giving oral presentations with charts, slides, maps, and models than at writing research papers, found it hard to follow the jargon and style in which many of the social scientists wrote. This meant that the staff had to spend valuable time translating the research findings, a task on which Heisel helped, before they could be worked into recommendations and proposals for consideration by the West End Task Force. But the way in which some of the social scientists worked proved a more serious difficulty than the way they wrote.

Social scientists are inclined by training to discover the solution to problems they pose for themselves. As a result, when assigned a problem by someone else, they tend to redefine it in ways that influence either the scale of the solution or the time it takes to find the solution. David Ames, for example, conducted an in-depth and time-consuming study of the problem of relocating businesses which recommended a solution reflecting his definition of the problem as a scholar, rather than as it might more appropriately have been defined for this planning process. As a result, he finished it late and out of sequence with other related studies being carried out by other team members. Thus, one of the most important papers produced on the project was not adequately incorporated into the planning proposals and recommendations we forwarded to the West End Task Force and the city.

We had a similar problem with educational planning. Lewis Bayles came up with two problems on which he wanted to take the lead in doing the research. The first was solving the problem of improving existing elementary schools and junior high schools, including the adaptation of the "satellite" school idea for preschoolers through fourth graders in the Queensgate II area. The second was adapting the "educational complex" idea (then more familiarly known as the "educational park" idea) to serve not only Queensgate II but also the entire basin area and some students throughout the school district. Bayles was clearly more excited about the latter, which he envisaged as involving the rehabilitation of some existing schools, the phasing out or relocation of others, and the use of specialized teaching staff and expensive equipment in special programs for the handicapped as well as in conventional schools. For him, the physical design, changes, and educational innovations associated with developing the educational complex plan seemed more exciting and more likely to have more significant influence in Cincinnati and other urban areas. Therefore, he wanted us to devote a major research effort to how the complex should be done.

Doubtless he was right about the potential significance of planning such a complex, but we could not go along with his desire to spend a great deal of time on it. In the first place, it would have taken too much time to do the research necessary for such a proposal. Beyond that, however, the complex site would have to be located outside our area of planning jurisdiction; its adoption would have required fundamental policy changes by the school board, and its execution would have been very expensive. These factors meant that such a plan would take a long time to carry out, and the longer it took, the greater the likelihood for changes and the less the chances for controlling the variables involved. We decided it safer and more prudent to give top priority to the satellite school idea and to spend less time on the educational complex, which appeared in our final report in preliminary form as "a future possibility" for "future consideration."

The prevailing mode of social scientific research, particularly the tendency for scholars to work on their own problems to be approached from their own discipline, created other difficulties for us. We would have preferred, for example, that the team work as a team on particular problems without reference to the academic discipline of team members. That never happened, but we did manage to integrate architects or planners with one or another of the social scientists, as we did in the educational planning. In addition, an architect teamed with the psychologist to survey the preferences and behavior patterns of West End residents with respect to housing, recreation, shopping, and other aspects of neighborhood life. And two planners and an architect did the early and basic work on housing schemes and neighborhood design. But only one instance of cooperation between social scientists took place. The psychologist and the geographer jointly undertook an attitude survey of local businessmen.

We even had difficulty in keeping touch with the social scientists, who tended to go off with their research assistants without reporting regularly to the office on what they were doing, how they were doing it, or what they were finding. This complicated our efforts to integrate the research and planning and forced us to constantly rearrange the sequence of the various components of the plan. Suspecting that the problem stemmed in part from some of the social scientists' lack of confidence in the director, who they seemed to respect as planner but not as an administrator or social scientist, we looked outside the team for someone to help improve our internal communications. Fortunately, the Vice President for Research at the time was Bob Carroll, a long-time faculty member at the university and a sociologist with a good reputation among his colleagues as both a social scientist and administrator. On two occasions he brought isolated social scientists into more effective communications with planners and architects.

Carroll also proved useful as a conciliator in instances of conflict between planners or architects and one or another of the social scientists. These disagreements stemmed usually from the social scientists' tendency to work alone rather than cooperatively, but they also came from another difference between the two groups. Social scientists are reluctant to make generalizations that to them or to their peers may seem to go beyond the reliability or the validity of their evidence. Academic planners resemble social scientists on this point, but, like professional planners outside the academy, most academic planners have had enough experience in planning to know that the time comes when decisions must be made, which often means that generalizations must be made even when information or knowledge is incomplete. Thus the social scientists balked for a long time at making any estimates of the possibility or probability of developing a plan which might influence middle-class families to live in an area close to the central business district, because they lacked the time and resources to make a scientific determination of the issue. This frustrated the planners and architects, and it infuriated some of them when some of the social scientists took to giving "lectures" in team meetings and informal conversations about the "professional responsibility" of academics, regardless of their discipline, to refrain from making generalizations and decisions on the basis of unscientific procedures.

The pace of the team's work was also slowed by worries about other potential and actual conflicts stemming from the actions of team members within the university. For example, planners on the team invited community activists associated with Queensgate II to speak in their classes and members of the faculty research group used data from and information about the project as instructional material. This was a major benefit of the partnership for the university, of course, but we feared the press might pick up and publicize stories about an activist or one of the social scientists criticizing the way the city or university was running the program or the social implications of a team recommendation to the Task Force. We also had a short but nonetheless distracting argument about the danger that faculty might use Queensgate II project data in publications which could antagonize other participants and jeopardize the outcome of the planning or implementation process, the latter of which we then felt might take several years.

But the major problem within the university was the split between academics and student assistants on the project and what they called "the administration." Both the faculty and students tended to be politically liberal and sympathetic to various militant or liberal perspectives on the position of minorities in American society. As a consequence of those views and of the research done on the project which supported some of those views, faculty and students tended to side with West End blacks or their supporters in controversies between the city and West End residents and black activists. Most university administrators who became involved in the project did not share the faculty members' and students' political and social views, and therefore tended to line up with the city in such controversies. This tendency was reinforced by the fact that the university, as a municipal institution, received financial support from the city by means of taxes levied by and administered through the city under an arrangement by which the mayor appointed more than half the members of the university's board of directors. Thus relations between the team and the university administration were strained throughout the process, particularly toward the end, when some black residents of the West End and some members of the Task Force sought to enroll team members on their side to break or overcome what seemed to them an alliance between the city's urban development staff and the university administration on an aspect of the plan. At one point the split between academics and administrators became so severe that one team member believed his faculty position to be in jeopardy because of administration animosity.

We also had to spend time resolving conflicts stemming from the demeanor of team members in dealing with residents of the West End and their representatives. In the course of his field research, for example, one "liberal" faculty member aroused community resentment and complaints by pointing up problems in Queensgate II to residents, and another created problems for us by assuming an attitude of *noblesse oblige* toward residents. Residents also grew hostile toward the team because the two spoke different languages. That is, "density" meant number of persons per acre to team members, but "overcrowded" to residents. To the university team a "system" was a concept for dealing with the multifactored urban situation. To the residents, "the system" was a pejorative to describe a society and a political process which worked to the disadvantage of minorities. It was something "to beat." And social scientists used the phrase "lower class" as a neutral and descriptive analytic term, while many residents used the phrase interchangeably with "*low* class," which to them meant uncouth and degraded.

We also had to work with West End residents to allay distrust of the university team aroused by conflicts for which the team or its members were not responsible. Black leaders in the West End had grievances against City Hall arising from events stretching over a period of years, many of which stemmed from previous urban renewal activities. Indeed, throughout the 1960s these black leaders depicted City Hall as the symbol and chief agent of the insensitive, oppressive, and antagonistic Cincinnati "establishment," and they grew increasingly militant in response to many city projects and their allegedly racist origins and consequences. Gradually, members of the city administration grew impatient with these outbursts of militancy, and by the time our project began many were furious at the behavior of the militant blacks. This attitude hardened in the middle of our project, when William Wichman was succeeded as City Manager by Richard Krabach, a much more conservative, authoritarian, and "hard-nosed" administrator. His less than charitable attitude toward militant blacks in general, and toward community participation in particular, led to a boycott of the planning process by blacks on the West End Task Force which stopped work on the project for six months. Meanwhile, blacks and the city argued over the role of the Task Force, during which each side either accused or suspected the university team of siding with the other and black leaders' doubts about us and the university team reached their high point.

The long-standing animosity of black activists toward the public school administration also hindered our educational planning, which required that we develop cooperative relationships with central administrative figures in the school system. Black activists believed that the two associate superintendents, rather than the superintendent, constituted the real powers in the school system through a chain of loyalties in the bureaucracy and among principals and teachers built by a process of appointing over the years disproportionately large numbers of Irish and German ethnics to school posts. To the black militants, this ethnocentric and nepotistic system meant that blacks got fewer and less desirable instructional and administrative posts in the sytem than they deserved, and it explained why the single black high in the administrative ranks occupied a "token" and dead-end position of little influence. The militants, in short, did not trust the school bureaucracy, and they suspected anyone who supported its policies of sharing its allegedly racist character. Therefore, we and the team members had to be careful of what we said about school affairs and avoid appearing to be too close to a bureaucracy with which, for planning purposes, we had to work out relations of mutual trust and cooperation.

On one occasion, however, we ran into trouble with West End residents on what was to both sides a very important policy question. Early in the process the team recommended the construction in Queensgate II of multiunit housing town houses. In a series of closed, crowded, and tense meetings, residents of the area made it clear to us that they wanted nothing less than single-family detached houses with white picket fences around all four sides. The team countered that the small amount and high cost of land in Queensgate II ruled out detached single-family dwellings, a position which angered the blacks, who regarded such housing in the inner city as a dream to which they were entitled, just as middle-income whites had fulfilled such a dream in the suburbs.

The final category of conflict with which we had to deal pitted the university team against the city, specifically against people in the city's Urban Development Department. The first of these occurred when the urban development director and his staff proposed making Queensgate II into a massive new town-in town complex of the kind and scale recently undertaken in New York, Chicago, and Boston. The university team opposed this notion, arguing that it did not square with the social and cultural ambiance of Cincinnati and therefore would not work. We preferred a more modest development, and after heated arguments, highlighted by threats to quit from our side and threats to get someone else to do the planning from the other, the issue was resolved.

The other major conflict with the urban development department took place over the choice of an economic consultant that the university team wanted hired to do a marketability study of housing and related developments. Urban Development encouraged us to consider Real Estate Research Corporation (RERC) of Chicago, which we learned later had signed many contracts concerning urban developments with various city departments. Urban Development Department personnel even arranged for meetings and correspondence between RERC and team members. Thus, before we had gathered information on other possible bidders for the consultantship, RERC staff had met with the team on three occasions in Cincinnati, team representatives had flown to Chicago for another meeting, and RERC staff had drafted papers about the Queensgate II project.

Nonetheless we insisted not only on taking other bids but also on taking them seriously. The list we finally assembled consisted of RERC, Robert Gladstone Associates of Washington, D.C., and Barton-Aschmann Associates of Chicago. Meanwhile, the Urban Development Department placed mounting pressure on us to hire RERC. Much to the department's chagrin, however, we selected Robert Gladstone Associates, a firm frequently and enthusiastically recommended by people with whom we discussed the matter and a firm with a strong national reputation and abundant experience in doing urban economic analyses. Clearly this was an awkward situation (to put it mildly), and it was weeks after the incident before the tensions subsided.

The difficulties of integrating social science research with planning and the conflicts within the university and between the team and the city thus prevented the university from reaping all the benefits many of us thought it might reap from the partnership. Many of us hoped it would give faculty and students real-world experience valuable to them as teachers, scholars, and as practicing architects and planners. That much was surely accomplished. But many also hoped the project would tighten and improve the university's relationship with City Hall and the black community. That, in our judgment, did not happen.

Nonetheless, the university team successfully completed its job as we saw it. The team developed proposals and recommendations which the West

End Task Force shaped into a plan acceptable to the City Planning Commission and City Council. That plan bears clear marks of the influence of both the citizens and the social scientists who participated in the process. And the plan broke with conventional planning wisdom by proposing modest-scale residential development for a heterogeneous population on the central business fringe. Though this plan may never be carried out, whatever is finally done in the Queensgate II area will be different from what anyone expected before this experiment with partnership planning in Cincinnati.

Postscript

We have tried to make clear in the general introduction and chapter headnotes our disinterest in this book with the problem of what happened to and in Queensgate II and the Cincinnati area after the completion of the Queensgate II planning process. Nonetheless, some of the reviewers of the manuscript for this book felt curious about those kinds of questions, and asked particularly why the plan has not been carried out. A satisfying response to that query would take more pages and time than either we or the publishers of this book can afford. But for readers who may be curious about this issue, and if such readers will remember that the Queensgate II plan or important aspects of it may yet be carried out, we can offer the following thoughts.

The Queensgate II plan, completed and adopted in 1970, has not been carried out because there has not been money to do it, a consideration few thought likely in the beginning. Queensgate II planning began in 1968, the year of the passage of the Housing and Urban Redevelopment Act of 1968, the federal legislation from which subsidies for developing Queensgate II were expected to derive. The 1968 act proposed a massive increase in the scale of federal assistance for low- and moderate-income housing by setting as a target the construction of 6 million such units over a 10-year period,[1] a stunning figure in light of the fact that fewer than 700,000 such units had been produced in the previous 30 years.[2] In addition, the act, through its Neighborhood Development Program, aimed specifically to encourage and hasten the completion of urban renewal projects, like that envisioned for Queensgate II, in big cities across the country. The act also signaled that federal support for the coordination of physical and social planning, started earlier in the 1960s, would be projected into the 1970s. The Queensgate II plan, adopted just 2 years after the adoption of the 1968 housing act, fit the pattern of the act's promise.

For a variety of now familiar reasons, the promise of the 1968 housing legislation was never met. First, President Lyndon B. Johnson, the driving force behind the act and the social policies it partially embodied, chose not to run for reelection in the fall of 1968, due largely to the controversy over the Vietnam war. Second, Johnson's successor, Richard M. Nixon, wanted to revise or curtail many of the Johnson Administration's programs, including some in the housing field. Third, the growing concern with increasing rates of inflation shifted domestic priorities and fired a mood of fiscal conservatism in the country which persisted throughout the 1970s and into the 1980s.

These circumstances caused trouble for the Queensgate II program as early as 1969, when George Romney, President Nixon's Secretary of Housing and Urban Development (HUD), undertook the implementation of the Housing and Redevelopment Act of 1968. Under previous urban renewal legislation, the federal government set aside and held in reserve sufficient funds for the completion of a federally approved project, regardless of the time needed for completion of the project. The Neighborhood Development Program, however, provided loan and grant funding only for urban renewal project activities which could be planned and carried out within a given fiscal year.[3] During 1969, Romney both insisted on a rigid adherence to this schedule and sought to reduce federal spending on urban development programs. As a consequence, Cincinnati's Urban Development Department spent federal money previously budgeted for Queensgate II in the Over the Rhine district and elsewhere, for city officials could not guarantee to the federal government the completion of a specific set of activities in Queensgate II during fiscal 1970, because in 1969 the policy plan had yet to be adopted and the specifications for the implementation of that plan remained incomplete.

Thus development activities in Queensgate II could not start until 1971, and by that time the Nixon administration had issued a signal that federal urban programs in the future would shift from the reconstruction of inner-city residential neighborhoods to policies promoting the dispersal of big city populations and, presumably, of the big city poor. The signal came in the form of Title VII of the Urban Growth and New Community Development Act of 1970, which aimed to encourage the creation of moderate-sized new cities beyond the crowded metropolises.

A corollary and confirmation of this dispersal approach to urban housing and social problems appeared in the Housing and Community Development Act of 1974. This legislation not only supplemented the decentralizing ideology of the Nixon Administration but also provided less federal financial aid to big cities, less in the way of federal leadership (out of deference to local control), and less of a federal role in encouraging the coordination of physical and social planning. Specifically, Section 8 of the 1974 act set up a "guaranteed rent" program authorizing HUD to contract with housing owners to pay the difference between the "fair market rent" of a unit and roughly 20 percent of the income of poverty-level occupants, and without reference to or preference for the location of the housing. Three assumptions underlay this program: first, that the poor should be encouraged to disperse across the metropolitan landscape rather than allowed (or forced) to huddle in big city slums; second, that poverty, not an inadequate housing supply, constituted the fundamental problem, and therefore subsidizing income rather than housing constituted an appropriate solution; third, that with subsidized incomes, the poor would prefer to behave like other Americans by moving out of the inner city, and therefore the private market, not the government, could and would "naturally" supply the needed housing in outlying neighborhoods, traditionally the most cherished building sites of private builders.[4] In short, the 1974 legislation combined a favorite nostrum of some liberal academicians in the 1960s (the trouble with the poor is that they are poor) with the classical libertarian bias of the 1970s,[5] which preferred market solutions rather than governmental solutions to social problems, into an income maintenance program designed to "aid the person, not the bricks,"[6] and to disperse the poor more thinly through urban residential neighborhoods.

Whatever the merits of this program, it did not "fit" the assumptions underlying the Queensgate II plan, completed four years earlier and in the spirit of the 1968 housing act. The Queensgate II scheme rested on the premise that the federal government would provide new housing in a particular area, not housing subsidies for individuals to secure existing housing anywhere in the metropolitan area, and on the notion that the poor and others might prefer or be persuaded to live in an inner-city area. Since the moment of its adoption in 1970, then, the Queensgate II plan was undermined by

Nixon Administration housing policies, and since 1974 no federal administration has devised programs sufficiently endowed to fund renewal projects or other urban programs of the kind or on the scale of development envisaged by the plan, or for which the Queensgate II project as defined in the plan would qualify.

Nor could Cincinnati's municipal government itself command the resources to do the job. This shortcoming seems to us not to have been due to a lack of will, however. To be sure, the preoccupation with racial and housing crises among the city's politicians and planners which provided much of the impetus for the Queensgate II planning process diminished in the 1970s. Yet in 1971 a coalition of council candidates fielded by the Democratic Party and the Charter Committee, an independent political organization which has played a major role in Cincinnati politics since the mid-1920s, swept into power in part by playing on pro-neighborhood as opposed to pro-downtown development sentiments, and in part by organizing Cincinnati's black citizens into a unified and steady voting bloc behind the Democratic-Charterite coalition. That coalition commanded a majority between 1971 and 1981, and such a coalition might legitimately have been expected to "deliver" to its friends, including those blacks and others who also formed part of the Queensgate II constituency.

Several facts of urban political life since 1971 may account for the coalition's failure to deliver. First, not many voters lived in Queensgate II. Second, most black voters in the 1970s, as now, lived in other parts of the West End or in the new hilltop black ghetto created by the renewal projects of the 1950s and 1960s. Third, and with the notable exception of William Mallory, most influential black politicians and officeholders in both the Democratic and Charter organizations have represented and/or lived in that new black ghetto or integrated neighborhoods. Fourth, city council members were (and are) elected at large, rather than by districts. Fifth, and probably most important, Cincinnati, like most American municipalities and especially the core cities of metropolitan areas, has been caught in a vicious fiscal crunch created by a declining tax base and inflation-driven expenditures, and compounded more recently by the Reagan Administration's determination to slash federal domestic expenditures. Under these circumstances, the coalition tried for a time to encourage both downtown and neighborhood development, but finally, in the last half of the 1970s, sent out the word that neighborhoods would henceforth have to do more for themselves. That, when combined with changing federal programs and priorities, left Queensgate II to pull itself up by its own bootstraps when, in effect, it had no bootstraps.

Nonetheless, some development has occurred in the Queensgate II area since 1970. The site reserved in the plan for a Town Center has been occupied by the Crosley Communication Center, a new facility which now houses WCET-TV, a public broadcast station; WGUC, the University of Cincinnati radio station; WRRM, an FM radio station; and a parking garage, although a smaller one than proposed for the Town Center in the plan. Second, a sheltered pedestrian bridge across Central Parkway linking the Crosley Communication Center and Music Hall has been built. Third, the Union Baptist Church has built a high-rise apartment complex, including housing for the elderly, a project conceived before the Queensgate II planning process but which the plan incorporated. Fourth, WEDCO/Mid-Cities has constructed a high-rise, moderate-income development that includes a plaza with adjoining space for offices and stores, and which has attracted a private commercial college as tenant but which otherwise has experienced difficulties in finding human and commercial tenants. Fifth, a new school, Taft Vocational School, has been built onto the Taft Senior High School. Sixth, and most recently, the Automatic Data Processing Company has moved its plant from its former location in Walnut Hills into the Queensgate II area. These buildings, to be sure, represent a moderate record of achievement, but, except for the Automatic Data Processing Company plant, they deviate only marginally from the substance and spirit of the plan.

Meanwhile, however, the city's Department of Development, formerly the Department of Urban Development, had by 1981 worked up some proposals for Queensgate II which ran counter to both the spirit and the substance of the plan. As we understand those proposals, they divide Queensgate II into four quadrants.[7] They proposed to put in one quadrant, an area bounded by Ezzard Charles Drive, Court Street, Central Avenue, and Central Parkway, a Park Department equipment maintenance facility, and to locate it near Central Avenue adjacent to the Crosley Communications Center. This project was not part of the Queensgate II plan, and would reduce the amount of space dedicated to housing in the plan.

Another quadrant, bounded by Ezzard Charles Drive, Court Street, Central Avenue, and I-75 Expressway, which contains public housing built in the 1930s as well as nineteenth-century residential structures, has been scheduled by Urban Development for residential treatment under the city's new division of Neighborhood Housing and Conservation. This scheme conforms generally with the Queensgate II plan, although it proposes the rehabilitation of the old housing stock along Clark, Hopkins, Elizabeth Streets, and Ezzard Charles Drive, a minor irony because the Queensgate II planners recommended rehabilitation of houses on Clark Street but considered the buildings on the other streets and on Ezzard Charles to be so deteriorated that rehabilitation would not be feasible.

A third quadrant, bounded by Ninth Street, "about" Seventh Street, Central Avenue, and I-75 Expressway, is slated for the development of "light" industry. This area already contains the recently constructed Automatic Data Processing Company's building, itself a clear departure from the Queensgate II plan, which prescribed for this territory a variety of kinds of low-rise housing. In addition, as of this writing, City Council is engaged in negotiations which may locate either an auto dealership or an extension of the city's convention center in this quadrant, where an old department store now stands near City Hall on the west side of Central Avenue, both of which would be incompatible with the plan.

The fourth quadrant, bounded by Court Street, Seventh Street, Central Avenue, and Plum and Race Streets (on the east), has been designated by Urban Development as an "acquisition area" for the construction of new buildings for a variety of uses. One proposal under consideration is the building of a high-rise apartment next to the Automatic Data Processing structure. These possibilities also violate the Queensgate II plan, which suggested housing, none of it high-rise, for the area.

Clearly, the Department of Development's consideration of moving Queensgate II in these directions, if realized, would substantially alter the Queensgate II plan. Yet prospects in the 1980s for the implementation of the Queensgate II plan, or critical aspects of it, may not be quite so grim. The City of Cincinnati seems committed to bolstering its tax base, if for no other reason than as a hedge against the continuing decline of federal (and state) assistance. Since it lacks territory within the municipality to accommodate the large and horizontal commercial and industrial facilities apparently preferred by private sector investors, it has adopted an economic policy centered on attracting smaller or less space-intensive businesses and restoring the city's old housing stock and its attractiveness as a place in which to live, a policy underscored in 1980 by the creation of a Historic Conservation Board, and in 1981 by the creation of the new Department of Neighborhood Conservation and Development. That new policy seems also aimed at promoting hotel and convention activities, and office construction downtown, combined with a significant increase in residential housing downtown and in areas adjacent to downtown, in both new and restored structures. Thus in the 1980s a variety of people, including real estate developers and speculators, historic preservationists, conservationists, representatives of downtown interests, city officials, and neighborhood organizers are now thinking hard about and scheming over the problem of who will do what in the Over the Rhine district, the West End generally, and in Queensgate II in particular, which together now seem the periphery of downtown if not a significant component of residential downtown itself, rather than slums.

No one, of course, knows how this thinking and scheming will turn out. It could result in the massive displacement of the poor from these three areas, but at this writing some neighborhood organizations, politicians, members of City Council, and city officials, especially in the Department of City Planning, on the Historic Conservation Board and among its staff, and in the Department of Neighborhood Conservation and Development, are supporting an intensive effort to prevent that outcome. Yet it may be that a major irony of the persisting Queensgate II planning experience will consist in the application of *some* solutions based on the conventional wisdom of the late 1960s (mixed-income and racially integrated housing, for example) to the solution of a new set of problems based on the conventional wisdom of the late 1970s and early 1980s. As we have suggested, however, that is another story, the subject of some other book.

NOTES

1. Bernard Frieden, "New Roles in Social Policy Planning," in Ernest Erber, ed., *Urban Planning in Transition* (New York: Grossman, 1970), p. 284.

2. Frieden, "New Roles in Social Policy Planning," p. 284.

3. William H. Claire, ed., *Handbook on Urban Planning* (New York: Van Nostrand Reinhold, 1973), p. 322.

4. Martin Mayer, *The Builders* (New York: W. W. Norton, 1978), p. 217. Extensive analysis of Section 8 is given by both Mayer (pp. 217-225) and Joel Friedman, Judith Kossy, and Mitt Regan, "Working Within the State: The Role of the Progressive Planner," in Pierre Clavel, John Forester, and William Goldsmith, eds., *Urban and Regional Planning in an Age of Austerity* (New York: Pergamon, 1980), pp. 262ff.

5. Mayer, *The Builders,* p. 217.

6. Mayer, *The Builders,* p. 217; referring to Raymond Barre.

7. Personal interview with a member of the staff, Department of Development, City Hall, September 15, 1981. Responsibility for the degree of accuracy in reporting the factual information that follows rests with the editors.

Appendix: General Goals and Objectives of the Queensgate II Plan

APPENDIX: General Goals and Objectives
of the Queensgate II Plan

Goals	Objectives
(1) Articulate community identity, in physical and social design.	(1) Establish physical and social focus of community life and activity (such as a "town center"), appropriately located.
(2) Functionally relate Queensgate II to adjacent and other areas.	(2) Design physical and social links between Queensgate II and rest of West End, Over the Rhine, and Central Business District.
(3) Improve quality of life of residents.	(3) Orient all design, planning, development, and programming to improvement of welfare and social and physical conditions of Queensgate II and West End residents.
(4) Improve and expand educational and economic opportunities for current and future residents.	(4) Expand elementary and junior high school facilities and develop a "magnet" (model) school and educational program. Attract middle-income families.
(5) Provide housing and community facilities and social services at high standards with creative physical and social design.	(5) Use best and latest techniques for residential and public facilities design, and provide best "fit" between design of homes and facilities, and local family and community needs.
(6) Strike a balance between a small neighborhood and a large, macrostructure new town-in town development.	(6) Population of about 6,000 in 2,000 housing units.
(7) Provide for economically and socially mixed population.	(7) Half of 2,000 housing units for families with children, half for individuals and families without children. Make for transition of mix of income levels across income dividing line, Central Avenue. Minimize reflection of income and class differences in architectural design, within marketing limits.

Appendix 201

Goals	Objectives
(8) Develop a high-standard black or predominantly black community.	(8) Add "showcase" quality to design of Town Center, housing, streets, and the like, and provide for top-flight social events and quality services.
(9) Provide both private and publicly subsidized housing.	(9) Develop private marketing programs. Use rent supplementation and similar programs in private housing developments. Utilize available (1970) public housing funds equivalent to 500 units.
(10) Improve recreation.	(10) Increase recreation facilities and programs by some 70 percent.
(11) Better education for residents.	(11) Adopt recognized quality educational programs and innovative techniques, technologies, and approaches.
(12) Improve job opportunities.	(12) Encourage development of city-proposed Liberty-Dalton industrial area (to provide jobs for West End residents).
(13) Minimize relocation discomforts.	(13) Eliminate off-site moves for (1970) relocatees by allowing all current residents to remain in area; give them housing priority.
(14) Achieve high residential area environmental quality: aesthetics and design, sound, air, and so on.	(14) Eliminate through-traffic in Queensgate II residential areas to reduce noise. Provide high architectural standards. Apply environmental health standards.

SOURCE: Prepared by Thomas H. Jenkins and Jayanta Chatterjee from Queensgate II plan documents for a visiting lecture on policy planning in the Department of City and Regional Planning, University of North Carolina, Chapel Hill, 1970.

About the Editors

ZANE L. MILLER received his Ph.D. in history from the University of Chicago in 1966. He is currently Professor of History, Codirector of the Center for Neighborhood and Community Studies in the Institute for Policy Research, and Curator of the Urban Studies Collection in the University Archives at the University of Cincinnati. He is a member of the City of Cincinnati's Historic Conservation Board, the Steering Committee of the Hamilton County Democratic Party, the Ohio Historic Site Preservation Board, the Ohio Committee for Public Programs in the Humanities, and the Editorial board of the *Journal of Urban History;* he is also the U.S. correspondent for the *Urban History Yearbook.* He is the author of essays, articles, and books on American urban history, including: *Boss Cox's Cincinnati; The Urbanization of Modern America; Clifton: Neighborhood and Community in an Urban Setting* (with Henry D. Shapiro); *Cincinnati's Music Hall* (with George Roth); *Physician to the West* (coedited with Henry D. Shapiro); and *Urban Professionals and the Future of the Metropolis* (coedited with Paula Dubeck). His most recent book is *Suburb: Neighborhood and Community in Forest Park, Ohio, 1935-1976.*

THOMAS H. JENKINS is Associate Professor of Planning and Sociology at the University of Cincinnati. He has been Visiting Associate Professor at Harvard and Visiting Lecturer at the Architectural Association in London. He has lectured at several other universities, including Cornell, University of North Carolina, George Washington University, and Boston University, and has had professional experience in housing and urban development in Chicago, Boston, and Cincinnati. He has recently edited a book, *Home and Family in the 1980s,* and published an article, "Reciprocal Relations: A Critical Assessment of Man-Environmental Design Studies" (*Intersection*). His other recent articles include "Apartment Communities: Hypothesis for Future Housing Form" (*Ekistics*) and "Anticipation of the Impact of Growth and the Impact of Anticipated Growth in a Small River Town" (Volume 4, *The Small City and Regional Community*). His graduate work in sociology was completed at the University of Chicago, and in city planning, at Harvard.